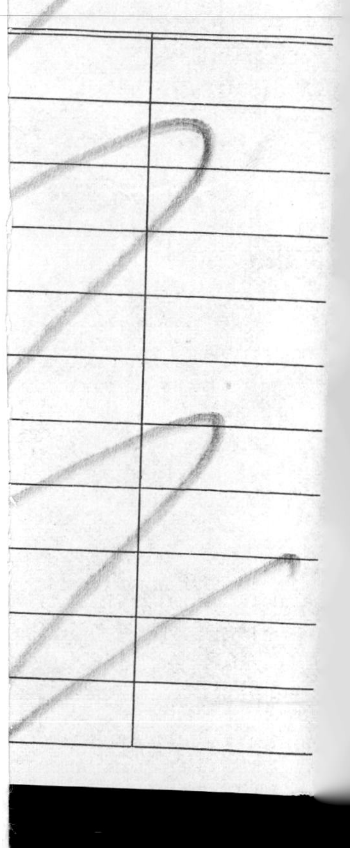

BIOLOGICAL STRUCTURE AND FUNCTION 3

THE INTEGUMENT

BIOLOGICAL STRUCTURE AND FUNCTION

1. *Biology of Bone*: N. M. HANCOX
2. *The Macrophage*: B. VERNON-ROBERTS
3. *The Integument*: R. I. C. SPEARMAN

THE INTEGUMENT

A Textbook of Skin Biology

R. I. C. SPEARMAN
B.SC., PH.D., F.I.BIOL.
Senior Lecturer in Dermatology
University College Hospital Medical School

CAMBRIDGE
AT THE UNIVERSITY PRESS
1973

Published by the Syndics of the Cambridge University Press
Bentley House, 200 Euston Road, London NW1 2DB
American Branch: 32 East 57th Street, New York, N.Y.10022

Library of Congress Catalogue Card Number: 72–88612

ISBN: 0 521 20048 2

Printed in Great Britain
at the University Printing House, Cambridge
(Brooke Crutchley, University Printer)

CONTENTS

Preface *page* vii
Acknowledgements vii
1 Introduction 1
2 A few methods useful in skin research 5

Part 1. Comparative morphology

3 Functional adaptation in the invertebrate integument 15
4 The integument of lower invertebrates 20
5 The integument of coelomate invertebrates 36
6 Invertebrate skin colouration 55
7 The fish integument 59
8 The skin of Amphibia 73
9 The skin of reptiles 83
10 The skin of birds 91
11 The mammalian dermis and subcutaneous tissue 101
12 The mammalian epidermis and its appendages 112

Part 2. Comparative functions

13 Thermal regulation 141
14 Chemical and neural control mechanisms 148
15 Comparative synthetic processes 159
16 Transport through the skin 170
17 Immunity 177
18 Development 181
19 General conclusions 189

Classification adopted: list of phyla 190
Further reading 191
References 191
Glossary 201
Index 203

Plates 1–12 are between pages 176 and 177

PREFACE

The purpose of this book is to provide in concise form a comparative account of the integument in both invertebrates and vertebrates, which has never before been attempted. It is directed towards honours students in the biological sciences, but the list of references will make it useful also to research students and others who wish to obtain a bird's eye view of the whole subject of Skin Biology.

The first part deals with the characteristics of the integuments of different animals, and the second part with comparative functions. Chapters can be read in any order.

The relative importance of pure and applied science is often discussed. Those who read this book will see that there are no real differences and that the one is a facet of the other.

It will be apparent to the reader that large gaps in knowledge still exist, which it is hoped will be a stimulus to further investigation.

In the reading list both elementary and advanced works are listed for readers from different disciplines. Where good reviews are available, these are listed in the bibliography in preference to individual papers. Elsewhere key papers are quoted. The reference lists in these works provide many important sources not listed in this book.

R. I. C. SPEARMAN
1973

ACKNOWLEDGEMENTS

I am grateful to Dr A. Jarrett, Dr P. A. Riley and Dr Mary Whitear for much helpful information, and also to all those who loaned photographs. I am also indebted to Mrs M. Henchoz for the way she has painstakingly typed and retyped the corrected manuscript.

1

INTRODUCTION

FUNCTIONAL IMPORTANCE

The integument or skin is one of the important organ systems of the body and forms a complete covering to the animal. To the outside is the external world with its highly variable and often hostile conditions, and within are the delicate living tissues with their stable physiological requirements. Forming as it does the boundary with the environment, it has a protective function. In consequence, in some species it produces a tough acellular cuticle or a mineralised shell. In others a surface covering of viscous mucus is secreted, or the outer cells die and form a durable keratinised layer. Sometimes a mineralised skeleton is laid down in the connective tissue region.

Sensory receptor cells as well as free nerve endings occur in the skin and supply the animal with information about the environment, but the requirements for protection and sensation are opposed, so that animals such as arthropods with a thick cuticle have most of their sensory nerves arranged to supply specialised sense organs, sensillae.

The integument is made up of an ectodermal epithelium, the epidermis, which lies on a bed of mesodermal connective tissue. In lower invertebrates, the latter grades into the general body mesenchyme, but in some higher invertebrates and in all vertebrates there is a distinct connective tissue layer usually with the body musculature at its inner boundary. Melanin and a variety of other pigments are responsible for skin colouration and produce camouflage patterns and sexual colour differences. Sometimes skin colour can be altered rapidly as a result of nervous or hormonal stimulation.

A feature peculiar to epithelia is their continued replacement in adult life from a germinal basal layer of cells which balances the loss of old effete cells from the surface. The rate of cell division is consequently much greater in epidermis than in internal organs, although less than in embryonic cleavage cells. However, nerve cells, also derived from embryonic ectoderm, never divide in the adult stage of higher animals. Clearly, the constant replacement of epidermal cells is a protective adaptation which prevents the integument from wearing away as a result of contact with the environment. The skin is a metabolically highly active organ.

HISTORICAL BACKGROUND

Aristotle of Samos in the fourth century BC noted integumental structures such as scales, feathers and hairs in his classification of animals. He also observed correctly that human male pattern baldness does not occur in eunuchs, although only in recent years has it been shown why this is so: that the male hormone testosterone causes a change in certain genetically predisposed hair follicles.

The Graeco-Roman anatomist, Galen of Pergamum, in the first century AD suggested that hairs and nails were made of the same substance, which we now know is keratin, but until the invention of the microscope in the seventeenth century serious study of the integument was not possible. Indeed, most anatomists until much later seem to have strangely regarded the skin as a hardened exudate from the blood.

The first comparative microscopical investigation of the integument was that of the Italian biologist, M. Malpighi (1628–94), who examined both invertebrate and vertebrate tissues through an early compound microscope and was the first to describe the structure of epidermis. Similar investigations were carried out by A. van Leeuwenhoek (1632–1723), an amateur microscopist of Delft, who made his own powerful single-lens instruments. The Danish anatomist, N. Stensen (1648–86) was the first to describe mammalian sweat glands. The observations of these early microscopists were not improved on until the nineteenth century. It is amazing how much topographical detail was discovered right up to the last decade of the last century with the aid of poor microscopes and mostly using teased preparations without any selective stains.

The first important advance in microscopy came with the introduction of achromatic lenses in 1824. A pioneer in the new comparative microscopy of tissues was the great physiologist Johannes Müller (1801–58), whose students in Berlin included T. Schwann, F. G. J. Henle and R. A. Kölliker. Schwann published his ideas on cells in 1839, later to be amplified by the pathologist R. L. C. Virchow (1832–1902): that tissues are composed of cells derived from other cells and their products. Henle, one of the best observers of his day, in 1837 was the first to show that epidermal cells are derived from a basal germinal layer. He also described the microstructure of hair follicles, and both he and Kölliker investigated nerve endings in the skin. Descriptions of the histology of both invertebrate and vertebrate integuments were given by Kölliker, G. Valentin, F. Leydig, C. Schmidt, W. C. Williamson and E. Haeckel during the middle part of the last century. C. Rouget in 1859 recognised chitin as a separate chemical constituent, and the hydrolytic products of chitin were later examined by G. Ledderhose. W. Biedermann between 1898 and 1926 made several important monographic contributions to both the invertebrate and vertebrate integuments.

In 1870, a German embryologist, the elder W. His, introduced the first microtome for serial sections, and by 1883, E. Abbé and K. Zeiss in Jena were producing good apochromatic oil immersion objectives. These two improvements led to a great expansion in the study of all animal tissues. L. Ranvier, P. Langerhans and W. Waldeyer during the last half of the nineteenth century re-examined the histology of mammalian skin, while the dermo-pathologist P. G. Unna was the first to investigate seriously the chemical nature of the mammalian stratum corneum.

During the last decade of the nineteenth century, with the introduction of improved staining methods, descriptions of integumental histology were published for most groups of vertebrates and invertebrates. Artifacts, produced by tissue fixation and processing and regarded by histologists of the time as genuine features of cells, were demonstrated by W. B. Hardy in 1899.

E. S. Goodrich in 1907 re-examined the dermal scales in living and extinct fishes and suggested probable evolutionary relationships.

The first quarter of the twentieth century saw the appearance of several monographic review works on comparative anatomy, and the establishment of histochemical methods for many cell constituents.

B. Bloch in 1917 introduced the Dopa reaction for melanin. This substance is oxidised enzymatically to melanin pigment by melanocytes. P. Masson in 1948 reviewed the subject of melanogenesis and suggested, as is now established, that melanocytes are glandular cells which secrete melanin and transfer it to epidermal cells.

Modern investigations on the insect cuticle were stimulated by the work of Wigglesworth, culminating in 1939 in his *Principles of Insect Physiology*.

Since 1940, light microscopic histochemistry has greatly expanded with the development of improved techniques, including those for enzymes. During the 1939–45 war, the phase contrast microscope was produced in Germany, and this period also saw the development of electron microscopy, with histological methods for tissue sections in the early 1950s. Biochemistry also leapt forward after 1940 with the introduction of improved methods of separation; in particular chromatography and the use of radioactive isotopes. The pioneer studies on protein molecular structure by W. T. Astbury and J. D. Bernal in the 1930s have grown now into the science of molecular biology. Histochemical methods for electron microscopy are being developed, and techniques for several enzymes have been published. This will make it possible to relate chemical localisation seen under the light microscope to ultrastructural differences visible only under the electron microscope.

ECONOMIC INTEREST IN THE INTEGUMENT

The study of wool, fur farming, leather which is tanned collagen, silk, toxic skin penetrants including insecticides, skin diseases and their treatment, and cosmetics are various economic aspects of what has come to be termed Skin Biology. Indeed, because of the greater research effort spurred on by necessity, more has been published about the integument of insects and of mammals than for any other animals, and little is known for some groups.

The skins of different invertebrates and of vertebrates show many similarities, but important differences also occur which will be discussed in the following chapters. In Part One, the characteristics in each group are dealt with, and in Part Two, comparative physiology is discussed. First some research methods are mentioned (Nordenskiöld 1946; Richards 1951; Montagna and Hu 1967; Van Abbé, Spearman and Jarrett 1969; Ryder and Stephenson 1968).

2

A FEW METHODS USEFUL IN SKIN RESEARCH

MICROSCOPICAL METHODS

MICROTOMY

Thin sections of skin can be cut on a microtome, 5–7 μm thick for light microscopy, and 0.01–0.05 μm thick for electron microscopy. Fresh, frozen or processed sections can be examined by phase contrast, dark ground illumination or polarised light. Frozen tissue is sectioned thinly in a cryostat. This is essential for enzyme histochemistry and consists of a deep freezer in which a microtome, usually of the simple rocker type, is placed and operated from outside the cabinet by cords and pulleys. The temperature in the cabinet is maintained at around −20 °C for most procedures. Fresh skin is best mounted in a mucilagenous material such as 5 per cent aqueous carboxymethyl cellulose, which when frozen by solid carbon dioxide or by a freezing aerosol spray hardens to a good consistency for cutting. Sections are mounted directly on cover glasses. A problem is free fat, which does not freeze hard enough for satisfactory sectioning so that adipose cells need to be frozen to a lower temperature.

Paraffin embedding methods are widely used for histological examination of skin. Water and lipoidal substances are extracted with solvents and are replaced by wax which permeates the tissue and can be readily sectioned when it sets hard. Prolonging the time in absolute alcohol or in clearing agents over-hardens the tissue, but dehydration and clearing can be continued in cedar wood oil after ethanol which helps to keep the skin soft. It is then cleared for a short time in benzene and taken through molten paraffin wax. The higher melting point 58 °C wax is useful for cutting tough specimens. Although a light rocker microtome is satisfactory for frozen cryostat sections, it is less good for paraffin material. A rotary microtome is usually satisfactory but very tough skin is best cut on a heavy base sledge microtome in which the knife is moved across the tissue.

Paraffin material is first fixed in a protein fixative. We have found that 70 per cent ethanol is surprisingly good and formalin is also widely used for fixation (Jarrett and Hardy 1957).

STAINING METHODS

Haematoxylin and eosin are the most valuable combination of histological stains for both invertebrates and vertebrates. Fluorescent stains are also useful (Jarrett and Spearman 1964). Mallory's triple staining technique is useful for insect cuticle (Noble-Nesbitt 1963): the endocuticle stains blue, the mesocuticle when present is red, and the refractile tanned exocuticle is unstained. Richards (1951) discusses various methods for arthropod cuticle. For general histological techniques see Lillie (1966). For hair morphology see Wildman (1954).

ARTIFACTS

It is important to realise that all fixation and processing methods produce artifacts of some kind. Thus, dermal connective tissue is grossly altered by paraffin methods due to the effects of protein fixatives on collagen.

HISTOCHEMISTRY

Substances detected in cells by histochemical methods may be bound to structural proteins as with cysteine and certain phospholipids, or free such as cysteine in glutathione, and fats. The distribution of structural chemical constituents is best examined in paraffin-processed tissue fixed in a narrow-spectrum protein fixative. Comparison can then be made with cryostat sections to show both bound and free material.

Histochemical methods for substances vary in their specificities. For histochemical methods see Bancroft (1967), Pearse (1968) and Thompson and Hunt (1966). One of the best methods is the dihydroxy-dinaphthyl-disulphide technique of Barrnett and Seligman for sulphydryl groups. Peracetic acid oxidation of cystine in keratin to cysteic acid followed by staining with thioflavine T gives a yellow fluorescence in cystine-rich cells. Only a few other substances are oxidised and cysteic acid is not a natural constituent. Thioflavine T also fluoresces with nucleic acids, and after initial treatment of sections with specific nucleases will confirm the distribution of deoxyribonucleic acid (DNA) and ribonucleic acid (RNA) (Jarrett and Spearman 1964). The acid haematin method is the best available for bound phospholipids. Free phospholipids can be extracted selectively in pyridine, but not protein-bound phospholipids.

The periodic acid Schiff (PAS) reaction stains red with certain carbohydrates and also with some proteins. Chitin, other neutral mucopolysaccharides and glycogen are positive. Specific enzyme digestion methods before the PAS reaction make the method more selective, but it is far from satisfactory. Acid mucopolysaccharides of connective tissue give a

weak reaction or are negative. Chitin can be demonstrated by the blue reaction in the diaphanol–iodine–zinc chloride method.

Hydrolytic enzymes in cells are demonstrated by two types of methods, and in both a product of the reaction is precipitated and visualised. For example, acid phosphatase hydrolyses the substrate sodium beta-glycerophosphate at pH 5 which releases phosphate ions. These then react with soluble lead nitrate in the incubation medium to give insoluble lead phosphate. The lead in the reaction product is sufficiently dense to be visualised by electron microscopy, but for the light microscope the lead phosphate must be converted to dark brown lead sulphide by treatment with ammonium sulphide.

In the dye coupling method, sodium alpha-naphthyl phosphate is hydrolysed and alpha-naphthol released, which couples with a diazonium salt in the mixture to give an insoluble coloured azo-dye.

Dehydrogenases remove hydrogen from substrates in tissue oxidations. They are demonstrated by tetrazolium salts which are water soluble, but when reduced by hydrogen are precipitated as insoluble coloured formazans.

SPECIAL METHODS

Both melanin and quinone-bonded cuticular proteins are bleached by hydrogen peroxide and also by peracetic acid, and are blackened by a weak solution of silver nitrate: the argentaffin reaction.

Cutaneous nerve endings for light microscopical examination may be stained by gold reduction or by the argentaffin reaction with alkaline-buffered silver nitrate. The lipid in myelinated nerves stains with osmium tetroxide, useful in electron microscopy. The enzyme acetylcholinesterase can be demonstrated at cholinergic motor nerve endings (see Weddell, Palmer and Pallie 1955).

Scales and teeth can be decalcified with minimum disturbance in an aqueous solution of the metal chelating agent, sodium ethylenediaminetetracetic acid (EDTA).

ULTRASTRUCTURE

For electron microscopy, tissue is fixed in buffered osmium tetroxide which gives good structural definition. However, osmium is a wide-spectrum fixative which reacts with unsaturated lipids as well as some protein groups. As a result, some soluble constituents in cells removed in paraffin techniques are precipitated in osmium-fixed tissue. Initial protein fixation in glutaraldehyde is required for most histochemical procedures or when specific solvent extractions are interposed between glutaraldehyde

and osmium post-fixation. Osmium and heavy metals are dense to the electron beam and so are useful as EM stains.

A valuable combination of EM stains for cellular structures is uranyl acetate and lead citrate. Phosphotungstic acid gives good contrast to protein fibrils, including collagen.

AUTORADIOGRAPHY

This is useful for the localisation of injected radio-active labelled substances in tissues. Radiation emitted from the isotope in the section impinges on a photographic emulsion coated onto the slide. The best localisation is observed with low-energy beta particles from tritium (see Ambrose and Easty 1970).

SURFACE PATTERNS

Plastic replica methods can be used to identify mineral crystal patterns in fractured surfaces of mollusc shells. Scanning electron microscopy can also be used to show the integumental surface topography. For electron microscopy see Kay (1965) and Meek (1970).

TISSUE CULTURE

The culture of skin cells is useful in the study of differentiation and the interaction of different types of cells. Combined with time-lapse cine photography, it is used to examine cell division, cell movement and formation of junctional complexes between adjacent living cells (see Ambrose and Easty 1970).

SOME BIOCHEMICAL AND BIOPHYSICAL PROCEDURES

CHEMICAL ANALYSIS

Hughes (1959) separated mite cuticle from other tissues by trypsin digestion and then hydrolysed the separated cuticles in NaOH or in HCl respectively under nitrogen to prevent autoxidation of cysteine to cystine. Amino acids were later examined by chromatography. Electrophoresis is also used.

Chitin can be detected by the chitosan reaction, which however cannot be relied on when results are negative (Richards 1951). The alpha and beta forms of chitin are determined by X-ray crystallography. Digestion by specific chitinases will demonstrate different types of chitin and is most useful.

Keratin can be broken down into sulphur-rich and sulphur-low fractions by either oxidation in peracetic acid and treatment with ammonia, or by reduction with thioglycolate and alkylation with iodoacetate. Gillespie (1965) finds the reduction method better for subsequent chromatography and electrophoresis of amino acids.

BIOPHYSICAL ANALYSIS

X-ray crystallography shows whether the alpha-helical or beta-pleated fibrous, or the non-orientated molecular, forms of biological substances occur. See Davson (1970, Alexander, Hudson and Earland (1963) and Elden (1971).

High-resolution electron microscopy can be used to determine the molecular structure of silk and the arrangement of keratin filaments in hairs, as well as for cuticular proteins.

Injected radio-active labelled substances can be examined in tissue homogenates by a Geiger counter for gamma radiation, or by a scintillation counter for beta particles. In the latter method a phosphor is mixed with the specimen and this gives off light when activated by the beta particles. The light emitted is measured by a photomultiplier.

SEPARATION OF EPIDERMIS

Separation of epidermis from dermis is necessary for biochemical examination of tissue homogenates. This is achieved by brief trypsin treatment (Medawar 1941), by a solution of calcium chloride (Riley 1967), or by heating to 60 °C.

Keratin and sclerotin both resist digestion by protease even after prolonged incubation.

EXPERIMENTAL METHODS

GLANDULAR SECRETIONS

Integumental glands are generally too small to be canularised and so the secretion for analysis has to be washed from the skin surface. Differences before and after stimulation by neural or hormonal means can then be determined. Mucous secretions as in frog skin can be collected in isotonic saline, and lipoidal secretions as in mammalian sebum in petroleum ether. Constituents are examined by chromatographic and other standard methods. Gland sizes before and after stimulation can be measured in histological sections by means of a planimeter.

EPIDERMAL CHANGES

Measurements of epidermal thickness and counts of the number of metaphase figures in dividing cells per unit area are also used to determine the actions of physiological or pharmacological substances. Control animals treated similarly to the experimental animals except for the absence of the agent under test provide the base line. The progress of thymidine-H^3-labelled nuclei up through the epidermis from the germinal layer, or movement of labelled substances bound in the cuticle, has been used to determine rates of cell movement or cuticular growth.

Hairs in albino mice or rats can be dyed in order to determine rates and sequences of fibre replacement.

SKIN PERMEABILITY

Various in-vivo and in-vitro methods have been used to measure movement of water, electrolytes and other substances across the skin. In-vitro techniques use the skin as a partition membrane, and changes in distribution of substances on either side of the skin can be determined, such as by radio-isotopes. Skin penetration is discussed by Tregear (1966).

SKIN GRAFTING

This is sometimes useful. Techniques for mammals are discussed by Billingham and Medawar (1951) and Ebling and Johnson (1961).

ANIMAL MODELS

Some animals are particularly useful for experimental studies. Thus, the last larval instar of the blowfly *Calliphora* does not lay down chitin in the cuticle, which is useful for studies on the other major constituent, quinone-bonded cuticular sclerotins. The high rate of production of silk by the silkmoth larva *Bombyx* and by certain spiders is useful for studies on protein synthesis. The house mouse *Mus musculus* tail epidermis undergoes contrasting types of keratinisation in the scale and hinge regions, and it can be used to determine the effects of substances which alter (modulate) keratinisation, such as vitamin A (retinol) (Jarrett and Spearman 1964). The sebaceous glands of the laboratory rat are widely used to study sebum secretion and have provided a model for human sebum secretion. The cavy ear is useful for melanocytes.

The function of sensory cutaneous nerves can be determined by changes in the pattern of electrical impulses after skin stimulation measured by electrodes placed on the nerves. Unfortunately, however, it is not practic-

able to measure electrical changes back as far as the receptor cells themselves by this method (Iggo 1968).

REGIONAL DIFFERENCES IN THE INTEGUMENT

All animals show inherited regional differences in morphology and composition of the skin. It is therefore important for the same area to be examined in control experiments. Differences sometimes occur between individuals within a species or in different inbred strains.

METHODOLOGY

When only the final effect of an agent such as a pharmacological substance thought to thicken the epidermis is known, epidermal thicknesses can be determined in a number of control and experimentally treated animals. The findings are then compared statistically and the degree of probability that the agent in question has an effect is assessed. Nothing, however, is proven by statistics, which are merely a means of overcoming the problem of induction: the inference of general principles from observation of particular instances, which is logically unsound. Occasionally the statistically improbable will turn up in experiments, which must be realised. Statistical analysis is not required when the mechanism of a process such as the Krebs tricarboxylic acid respiratory enzyme cycle has been worked out. In such cases, once repetition in a far smaller number of experiments has shown the mechanism to be valid, the findings are not invalidated by the discovery that the pentose pathway can be used instead. Neither is the mechanism of melanogenesis invalidated by the finding of albinism. This is the advantage of a good mechanistic hypothesis which may have to be modified, but is not necessarily toppled, by discovery of new information. Indeed, the new information should improve understanding of the mechanism. Experiments should be designed to refute hypotheses, as suggested by Popper (1963), and not merely to confirm cherished ideas. Eventually new information may require a complete reappraisal, as happened with the phlogiston theory of combustion, so that nothing is sacrosanct.

PART 1

COMPARATIVE MORPHOLOGY

3

FUNCTIONAL ADAPTATION IN THE INVERTEBRATE INTEGUMENT

CHARACTERISTICS

THE EPIDERMIS

In invertebrates this generally consists of a single layer of cells, sometimes fused to form a syncytium. In species with a thick cuticle the epidermis is often incorrectly termed the hypodermis. It undergoes two main types of specialisation. Sometimes it exosecretes a hard protective proteinous cuticle which may be strengthened with chitin or mineral salts, or a thick mineralised shell may be formed. Elsewhere the epidermis is usually ciliated and goblet cells secrete mucus. These types of epidermis often occur side by side in the same animal (Fig. 1).

Epidermal cells which produce a cuticle are columnar or cuboidal in shape and usually have the nucleus displaced towards the base of the cell. The exposed surface has a brush border, shown in electron micrographs as made up of microvilli (Fig. 2). In the epidermis these are associated with exosecretion, as distinct from the gut microvilli which increase the surface area for food absorption and in parasitic flatworm epidermis which resembles gut. Inactive epidermal cells, as occur in adult insects and under the shells of molluscs, are more flattened but remain capable of renewed secretory activity when repair is required.

In vertebrates and most invertebrates, the lower surface of the epidermis presents an even boundary with the underlying dermis, but in a few flatworms and in nemertines, individual epidermal cells dip down into the connective tissue.

THE SKIN CONNECTIVE TISSUE

The collagenous connective tissue beneath the epidermis is conveniently termed the dermis even in acoelomate animals where it is continuous with the general body connective tissue. Often in lower animals it is referred to as mesogloea or mesenchyme, but this is the same kind of tissue with the major constituent collagen in its various forms. Three important types of cells occur in the dermis: fibrocytes which exosecrete collagen and matrix

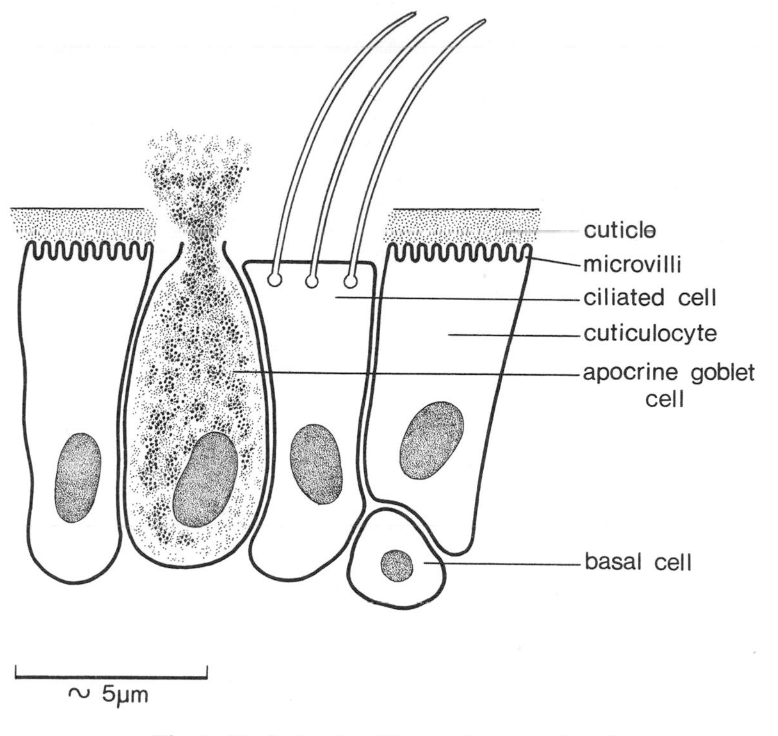

Fig. 1. Typical cells of invertebrate epidermis.

material, macrophages which are general phagocytes, and pigment cells. In life, the dermis behaves as a resilient elastic gel; largely a property of hydrated collagen. Elastin does not appear to occur in invertebrates although present in the vertebrate dermis. In some sponges and in echinoderms, a hard crystalline mineral skeleton is laid down in the connective tissue.

Immediately beneath the epidermis, there is generally a thin mucopolysaccharide-rich layer which under the electron microscope appears more electron dense. This is the basal lamina, a better term than basal membrane, as it is sometimes called. This is not a true membrane and migrant cells can move freely through it. A basal lamina is absent in sponges, coelenterates and a few turbellarians. In some other turbellarians, skeletal spicules are laid down in the basal lamina.

Muscles in the form of transverse and longitudinal bands for locomotory movement separate the integument from the general body connective tissue in more advanced invertebrates. In arthropods these muscles are attached to the integumental exoskeleton. Many acoelomates have a network of muscle cells in the dermis and thin sheets of muscle often occur

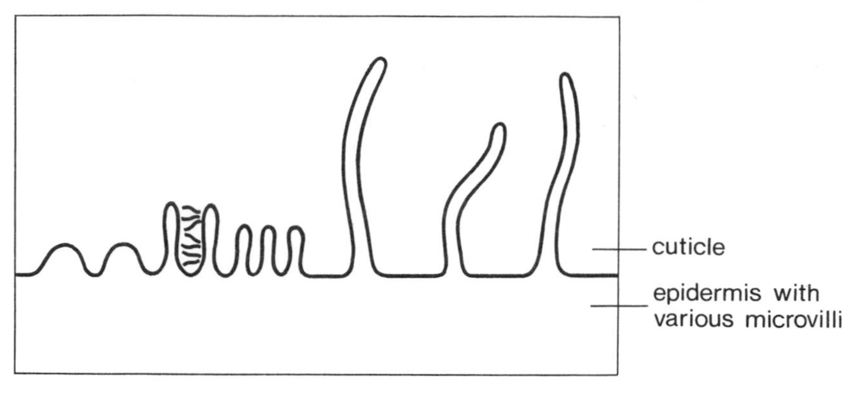

Fig. 2. Various secretory microvilli associated with the epidermal cuticle.

in higher animals. In a few turbellarians and in coelenterates the epidermal cells (epithelio-muscular cells) perform the function of body musculature through contractile filaments in their muscle tails. It is now realised that practically all animal cells contain some actin-type muscle filaments, important for cell movements such as phagocytosis.

In coelenterates, in some turbellarians, and in nemertine worms, a primitive nerve net occurs immediately beneath the epidermis, but in more advanced invertebrates, the nerve fibres are grouped together in superficial cutaneous nerve bundles.

FUNCTIONS

The invertebrate integument performs several important functions. Physical protection is afforded by epidermal exocellular proteinous or mucopolysaccharide secretions which either harden to form a cuticle or remain fluid as viscid mucus, and also by dermal skeletal elements. Higher vertebrates form a different type of epidermal covering layer in which the fibrous protein, keratin, is laid down within the outermost epidermal cells which then die.

The second function is the formation of a barrier between the external environment and internal organs to prevent inward or outward movement of water or tissue substances. In this respect most invertebrates have had very limited success. Except when a syncytium occurs, the epidermal cells are joined together by desmosomes with or without fusion of opposing plasma membranes. The various vertebrate-type junctions occur in insects (Smith 1968) and probably other invertebrates. In addition, the septate desmosome is peculiar to invertebrates (Fig. 3). This fused flange-shaped junction prevents water movement through the epidermal intercellular space (Locke 1965).

The third basic function is ciliary activity. Only small animals less than about three millimetres in length are able to swim or to crawl along the substratum by means of cilia. This is true of planktonic invertebrate larvae and of ciliate and flagellate protozoa. Among free-living turbellarians, the smallest flatworms use cilia for locomotion, but the larger species move by means of a muscular foot. Ability to reverse the beat is seen in small turbellarians, in sea anemones and some nemertine worms, and requires nervous control. Beat reversal occurs in the avoidance reaction of *Paramecium*. In larger flatworms and molluscs, where the cilia are used to transport food particles, to remove detritus or to produce surface water currents, reversal does not occur. Nervous control may be present or absent in different sites. Both activatory and inhibitory nerves occur in different molluscs. Denervated *Mytilus* gill cilia cease to beat one hour after severance of the branchial nerve to the area. Vertebrate ciliated epithelia are not controlled by nerves.

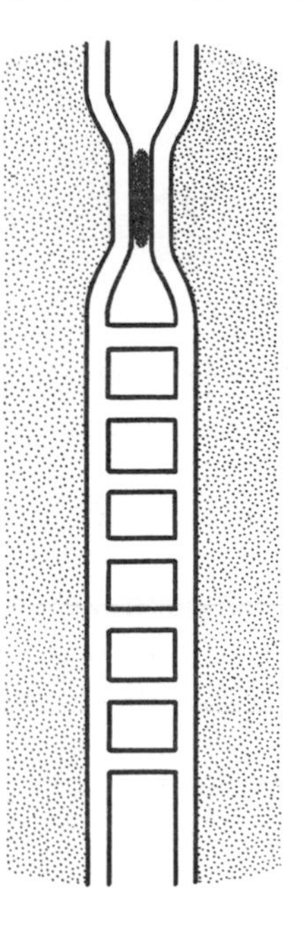

Fig. 3. A row of septate desmosomes between two invertebrate epidermal cells.

Cilia and flagella (Barrington 1967) are homologous structures which differ in their mode of beat. A flagellum, typified by that of a flagellate protozoan or of a sponge choanocyte, is much longer than a cilium and only one occurs for each cell, whereas several cilia occur over the surface of each cell. The beat of a flagellum is more complex as it functions as an independent unit. There are, however, gradations between the two.

The bending of a cilium or flagellum is caused by contraction of axial filaments of actin-type protein, similar to that found in muscle, actinomyosin. Inside each cell, fibrous strands connect the cilia with a single centriole or, in ciliate protozoa, with similar basal bodies (kinetosomes). In cross-section, cilia and flagella are bounded by a membrane. In all vertebrates and invertebrates examined, filaments (microtubules) inside are arranged in an axial, possibly skeletal, pair with a peripheral ring of nine, probably contractile, filaments (Fig. 4).

Movement of a cilium involves a rapid forward motion when filaments are contracted, followed by relaxation when the cilium returns to its former position by elasticity. Ciliated epithelia show metachronal rhythm, with each cilium out of phase with its neighbour in front by the

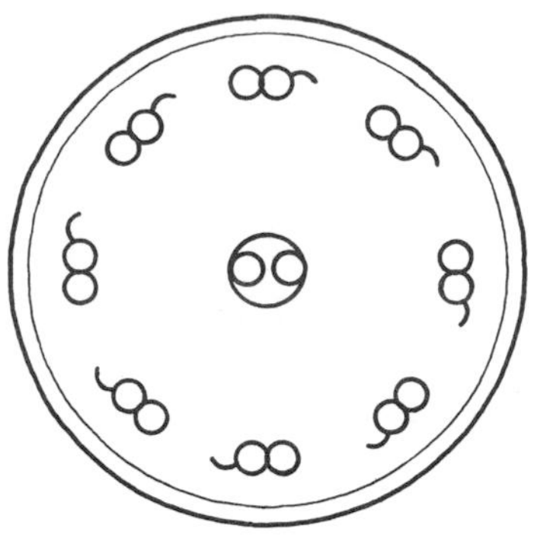

Fig. 4. Arrangement of microtubules in cross-section of a cilium (after Sleigh). (A ring of nine pairs instead of eight is now accepted).

same amount, but how this is coordinated has not been determined. As a result, waves of movements pass over the epithelial surface (Fig. 5). The action of cilia and their control have been reviewed by Riviera (1962) and Sleigh (1962).

The development of an extensive area of cilia of necessity precludes the formation of a hard cuticle or shell and thus reduces the protection afforded, but mucus, a watery solution containing mucopolysaccharides, is secreted and gives some protection.

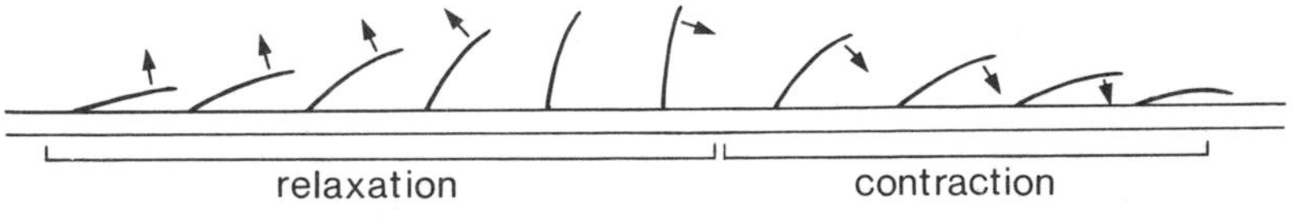

Fig. 5. Metachronal rhythm of ciliary movement.

The last, but not the least, important function of the invertebrate integument is neurosensation. In the lowest invertebrates which do not have specialised sense organs, the whole epithelium is irritable to environmental stimuli, but an early evolutionary development was the appearance of specialised sensory cells, probably derived from the epidermis, which act as receptors for touch, light and chemical stimuli. The lower portions of these cells are in close proximity with nerve endings (Carthy and Newell 1968; Steven 1963; Stern 1954).

4

THE INTEGUMENT OF LOWER INVERTEBRATES

PROTOZOA

The same basic functions of Metazoa are performed in Protozoa by the cell as a whole. Often there is a triple-layered plasma membrane 70–100 Å thick, which can be modified in various ways. The greatest complexity is reached in the ciliate pellicle, and in *Paramecium* electron microscopy has revealed a multi-membranous structure modelled around each cilium (Grimstone 1961). In surface view under the light microscope, the pellicle shows a hexagonal pattern of ridges containing trichocysts, with a single cilium in the centre of each hexagon. Trichocysts are membranous organelles associated in development with ciliary basal bodies and able to explosively discharge long threads, which temporarily anchor the animal or occasionally harpoon prey. Both skeletal and contractile fibrous proteins occur in Protozoan cytoplasm.

In the order Peritricha of the class Ciliata, the individual remains attached to the substratum for long periods by means of a basal stalk which is probably constructed of modified cilia bound together. In *Vorticella*, *Carchesium* and *Zoothamnium*, the stalk is contractile and contains actin filaments, but in *Epistylis* and *Opercularia*, this is not so, and the fibres contain sulphur-rich sclerotin. Electron micrographs suggest that the point of origin of the stalk is as an organelle for fibrous protein secretion (Pitelka 1963), comparable to the exosecretion of cuticular sclerotins by Metazoan epidermis.

Another resemblance to Metazoa is found in the resting cysts of fresh-water amoebae which are exosecreted through the plasma membrane and contain chitin (Kudo 1966). Many Protozoa secrete chitinised or mineralised tests, often with an elaborate architecture. Tests are probably always laid down in the cytoplasm beneath the plasma membrane. *Arcella* is an amoeba with a chitinous test. In the Foraminifera, the tests are usually strengthened by the crystalline calcite form of calcium carbonate. The millions of compressed tests from long-dead forams form the ooze on the deep ocean bed and also give rise to the chalk strata. In the Radiolaria, a silica test is formed, while cellulose is laid down in the plant-like Phytomastigina.

PORIFERA

Sponges present an evolutionary backwater. The swimming larvae are ciliated, but in the adult animal the body surface is covered with a layer of non-ciliated flattened amoeboid pinacocytes. These cells and flagellate choanocytes, which line the pore canals through which food particles are drawn, possibly represent the epidermis. The only other material present is a gelatinous connective tissue containing various amoeboid cells: sclerocytes which exosecrete skeletal fibres of spongin, a highly polymerised form of collagen, and others concerned with food digestion and with reproduction.

Spongin has a typical collagen X-ray diffraction pattern and is rich in glycine, hydroxyproline and proline. Two types of spongin are seen in electron micrographs: ultrafine fibrils 200 Å wide and broader fibrils 10–50 μm wide. The spacing of tropocollagen sub-units is the same as in vertebrate collagen (Florkin and Scheer 1968). The class Demospongiae contains the domestic bath sponge in which the skeleton is entirely of spongin.

Certain other sponges have in addition a mineralised skeleton. In the class Calcarea characteristic three-rayed spicules of calcite are exosecreted by sclerocytes and do not contain organic material.

In the class Hexactinellida, a skeleton of six-rayed spicules of silica is laid down inside the sclerocytes. The most elaborate development of a siliceous skeleton is seen in Venus's flower basket, *Euplectella*, in which there is formed a single brittle glass rod a metre or so in length but only a few millimetres in diameter; in addition there are smaller rayed spicules as in other species. Silica is not utilised by higher animals.

MESOZOA

Anatomically the simplest group of Metazoa is the Mesozoa, a minor group of small acoelomates with obscure phylogeny, all endoparasites of marine invertebrates. They are of interest because the body is made up of a ciliated epidermis, sometimes a syncytium, enclosing the reproductive cells with no other tissues present.

CNIDARIA

This phylum comprises the coelenterates with nematocysts. In these diploblastic animals the epidermis is a single cell layer over the mesogloea; the latter is sandwiched between the ectoderm and endoderm. The mesogloea has been isolated as a gelatinous sheet in the large Californian hydroid *Corymorpha*, and it contains collagen which is slightly different to that of mammals. For work on mesogloea see Florkin and Scheer (1968).

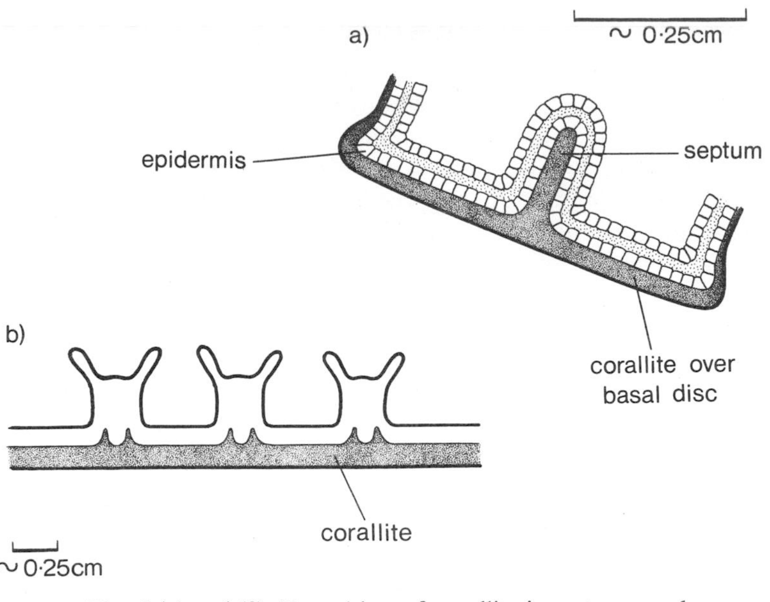

Fig. 6 (*a*) and (*b*). Deposition of corallite in a stone coral.

The basic structure of the epidermis is demonstrated in the fresh-water *Hydra* and is made up mainly of large epithelio-muscular cells. Mucus secretion by glandulo-muscular cells occurs especially in the pedal disc. Spindle-shaped sensory receptor cells, each with a projecting filament, are sandwiched between the epidermal cells and are in contact with endings from the nerve net. Germinal cells are wedged between the other epidermal cells. Adjacent cells are held together by rows of septate desmosomes which give a ladder-like appearance in electron micrographs (Lentz 1966).

The epidermis secretes a thin cuticle in *Hydra* except over the tentacles, which in both *Hydra* and sea anemones are ciliated. In the large floating Siphonophora, the sails of *Velella* and of the Portuguese man o'war *Physalia* contain beta chitin. In colonial Hydroidea such as *Obelia* the epidermis secretes a chitinous cuticular perisarc, and in the branched Hydrocorallina this is further hardened with calcite. The most extensive development of this type of calcified exoskeleton occurs in stony corals (Zoantharia) which form tropical reefs and atolls, and in the related solitary cup corals of temperate seas. In development a thick layer of corallite, mainly calcite in a thin proteinous matrix, is laid down over the pedal disc epidermis (Fig. 6). Radially arranged calcite septae are then formed between vertical invaginations of the disc epidermis, clearly seen in dried corallite.

The coral polyps concentrate mineral ions, mainly calcium, but also

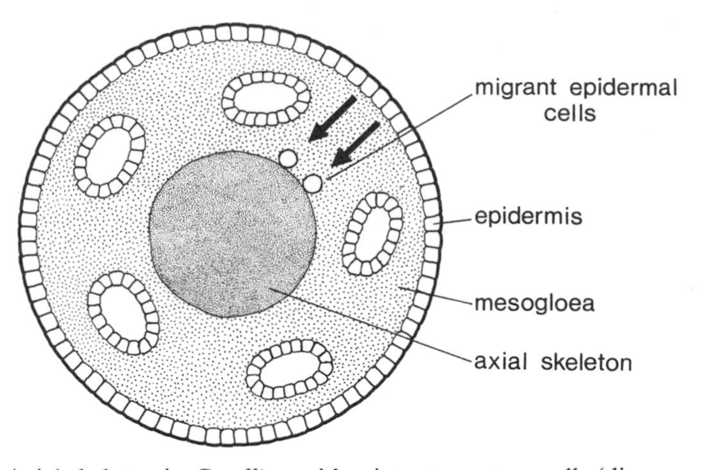

Fig. 7. Axial skeleton in *Corallium* with migrant secretory cells (diagrammatic cross-section).

strontium, from sea water. In the process of mineralisation, calcium together with carbonate ions diffuse out through the epidermis (Goreau 1963). Under alkaline conditions, calcium carbonate crystallises out, probably on the protein moiety previously secreted. Mineral deposition is helped by the presence of symbiotic algae. Thus, under laboratory conditions deposition of calcium carbonate was tenfold greater in the light than in the dark which, it was suggested, may be due to algal photosynthesis which produces a high local concentration of hexose sugars. The rapid oxidation of these sugars by the polyp epidermal cells with the formation of carbon dioxide would provide a local high concentration of carbonate ions in the surrounding water. The epidermal cells are rich in alkaline phosphatase, probably concerned with ionic transport across cell membranes which occurs during formation of the exoskeleton (Goreau 1963).

Beta chitin has been found in the calcified covering of some Hydrocorallina and Zoontharia, but not so far in the Alcyonaria. The latter sub-class includes the precious red coral *Corallium*, the organ pipe coral *Tubipora*, and blue coral *Heliopora*, together with the horny corals (Gorgonacea) and sea pens (Pennatulacea). In the Alcyonaria the axial skeleton appears internal although laid down by ectodermal cells. These cells migrate inwards from the epidermis in dead men's fingers *Alcyonium*, which has loosely arranged spicules of calcite in the mesogloea. In the red coral and organ-pipe coral similarly formed spicules are compacted together into a skeletal axis (Fig. 7). The blue coral differs from this in that calcium carbonate is in the aragonite form laid down as a thick deposit by an inner layer of ectoderm (Fig. 8). Probably this occurs by invagination of the basal disc epidermis which during growth burrows through the

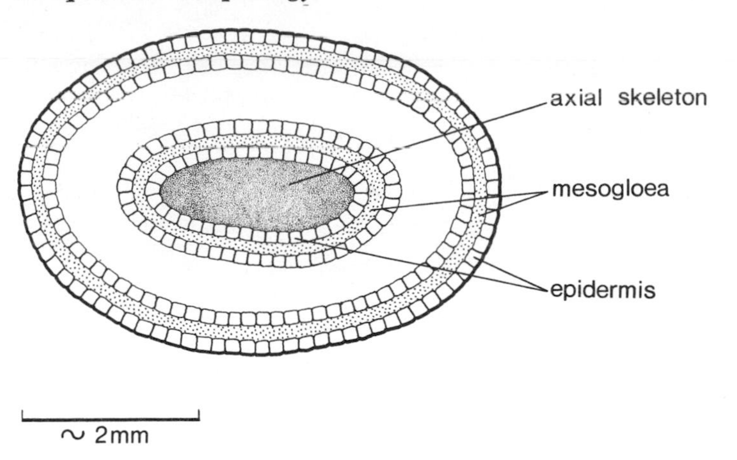

Fig. 8. Axial skeleton in the coral *Heliopora* (diagrammatic cross-section).

entire length of the colony. The axial skeleton is therefore really external as in Hydrocorallina and in the septae of stone coral. This is also true of the horny corals which have an axial skeleton of calcite spicules together with a peculiar horn-like structure named gorgonin rich in bromine, iodine and tyrosine. The sea pens have a particularly long horny axis.

In the black thorny corals (Ceriantipatharia) of the deep ocean floor there is a horn-like axis which contains beta chitin.

Nematocysts

The characteristic stinging organelles of Cnidaria are the nematocysts, which have a peculiarly involved development. They originate in germinal cells of the stomodeal epidermis in colonial species, in the medial body wall of solitary polyps, and in tentacular bullae of medusae. The immature nematoblasts then move through the mesogloea and endoderm into the gastric cavity. From there they migrate, probably by amoeboid movement, to the bases of the tentacles where they enter through the endoderm and end up between the tentacle epidermal cells with their trigger-like cnidocils directed outwards. In development, the nucleus of the nematoblast is displaced to one side and the cell is invaginated to give a sac. In the mature nematocyst this eventually occupies the larger part of the cell and houses the coiled stinging harpoon thread. The latter is a filamentous cytoplasmic protrusion, and in electron micrographs appears like a modified flagellum (Slautterback and Fawcett 1959). The remaining cytoplasm differentiates into a region with contractile filaments and a glandular region in which vacuoles of toxic injection fluid are formed. The sensory cnidocil, when touched by a small planktonic prey, triggers the contraction of muscle fibrils in the nematocyst which causes the explosive

discharge of the hollow thread through which poison is released. There is, however, considerable selectivity and a cnidocil does not respond to all physical stimuli, while some chemical stimuli produce a discharge. Chitinous barbs occur near the end of each thread. Stinging cells are used by coelenterates for food gathering rather than for defence. The toxic substances are mainly peptides and 5-hydroxytryptamine. The Portuguese man o'war, through the combined action of its numerous nematocysts, can produce a serious reaction in man due to release of histamine from human mast cells, but most species are only capable of paralysing small prey. Although nematocysts in general are not supplied with nerves since they act as independent organelles, an exception is in the pedal disc of certain sea anemones. Here, the nematocysts are under nervous control and their coordinated discharge, together with a mucous secretion, helps to anchor the animal (Ellis, Ross and Sutton 1969).

Marine turbellarians sometimes have functional nematocysts in their epidermis. The origin of these coelenterate stinging cells is the food which includes hydroids. When the rest of the prey is digested by the flatworm, the still-living nematocysts migrate through the gut wall and connective tissue into the epidermis of the new host as if it were another polyp.

CTENOPHORA

Because of their many morphological differences from the Cnidaria, the globular ctenophores are often classed as a separate phylum. The epidermis is either a single layer of vacuolate cuboidal cells or a syncytium. Often it is ciliated. Locomotion is by comb-like plates, ctenes, arranged in meridial rows. These are formed of laterally fused cilia, and so ctenophores are the largest animals to use ciliary locomotion. Each animal has two large tentacles which it uses to catch prey. The tentacle has a core of mesogloea lined by epidermis which contains specialised adhesive cells, colloblasts. These have a hemispherical head fastened to a coiled filament developed from the base of the cell, as in the nematocyst thread. When the thread is discharged, adhesive but non-toxic droplets on the head stick to and trap any passing prey (Hyman 1940; Florkin and Scheer 1968).

PLATYHELMINTHES

These acoelomate worms are divided into a free-living class, Turbellaria (flatworms), and two endoparasitic classes: Cestoidea (tapeworms) and Trematoda (flukes).

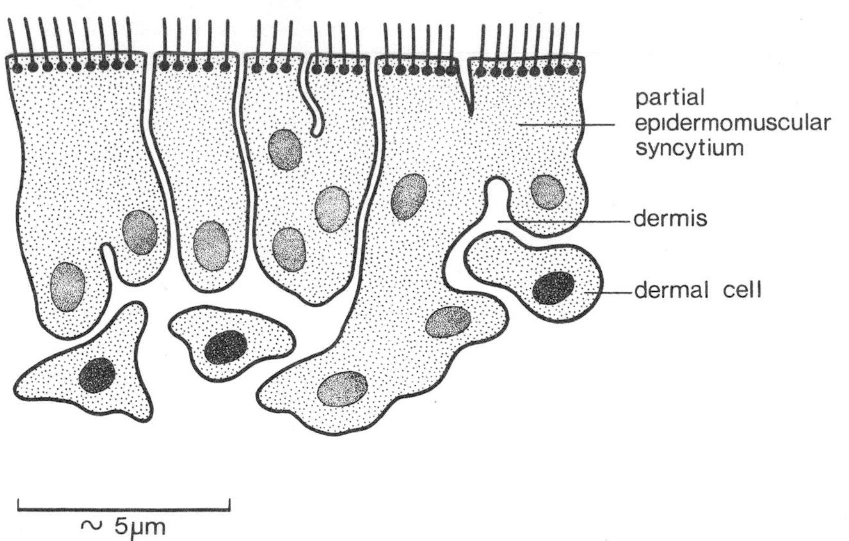

Fig. 9. Primitive epidermis in certain turbellarian Acoela, without a basal lamina (diagrammatic, after Beklemishev).

TURBELLARIA

This is a wide assemblage of flatworms of various sizes and degrees of complexity, of which the simplest is the order Acoela, containing species without a gut and with a dermal nerve net. In other orders a ventral nerve cord is developed. The most primitive epidermis is seen in the less highly organised Acoela. These interesting free-living marine flatworms, which are roughly oval in shape and less than 2 mm in length, are generally placed on the evolutionary tree close to the origin of the Metazoa (Hyman 1940; Hadzi 1963; Beklemishev 1969).

The most primitive type of epidermis is found in the Acoelate *Oligochoerus*. In this animal the exposed surface of the epidermis presents a smooth ciliated layer with some of the cells fused together, but the basal ends of the cells remain separate and project down into the connective tissue which has no basal lamina, although one occurs in more highly organised turbellarians (Figs. 9 and 10).

Both epithelio-muscular cells and true muscle cells occur in the free-living flatworm integument. Sensory receptor cells are wedged between the musculo-epithelial cells as in coelenterates.

Epithelio-glandular cells of turbellarians secrete peculiar rod-shaped bodies (rhabdites) which are stored in the cytoplasm and are released into the surrounding water only when the skin is irritated. They then swell and form a slime which covers the skin surface. Rhabdites probably trap and

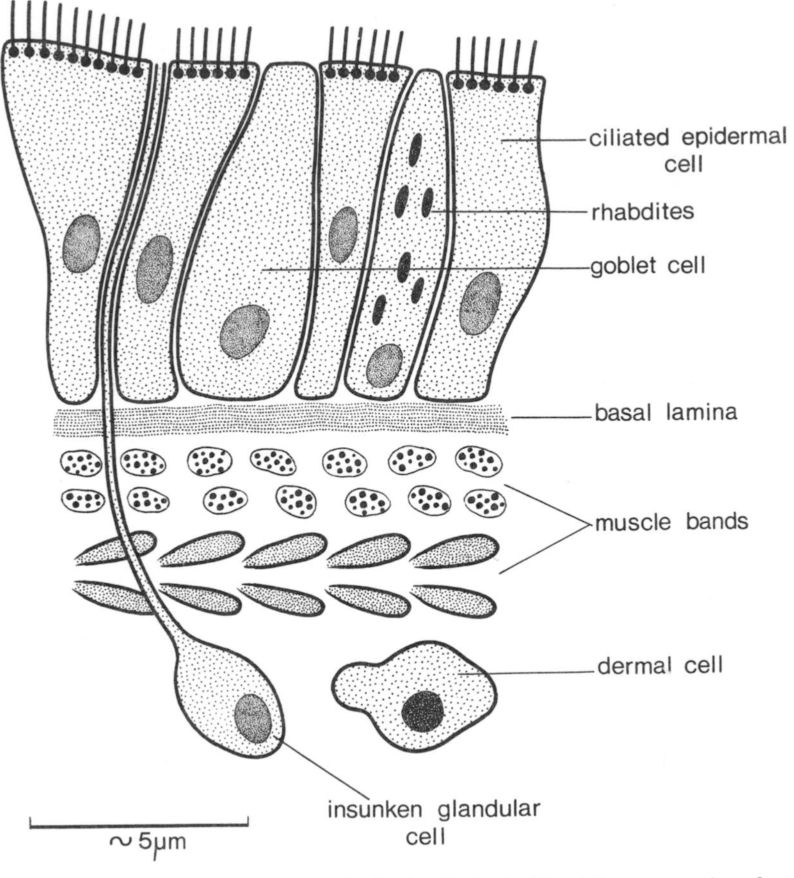

Fig. 10. Epidermis in certain turbellarian Polycladida (diagrammatic, after Beklemishev).

remove injurious foreign bodies and parasites from the surface of the worm (Hyman 1951*a*).

Types of Epidermal Organisation

Beklemishev (1969) derives four principal types of epidermal organisation from the primitive conditions in *Oligochoerus*. These are: (1) as occurs in many free-living turbellarians and in parasitic platyhelminthes; an epidermal syncytium with insunken nuclei hanging in the dermis and attached by narrow stalks of cytoplasm (Fig. 11). (2) The nemertine type of epidermis with a flat ciliated surface made up of cells with separate tapering ends in the dermis (Fig. 12), glandular and sensory cells being packed between the tapering processes. This latter type of epidermis also occurs in some turbellarians as well as in the enteropneusts, such as

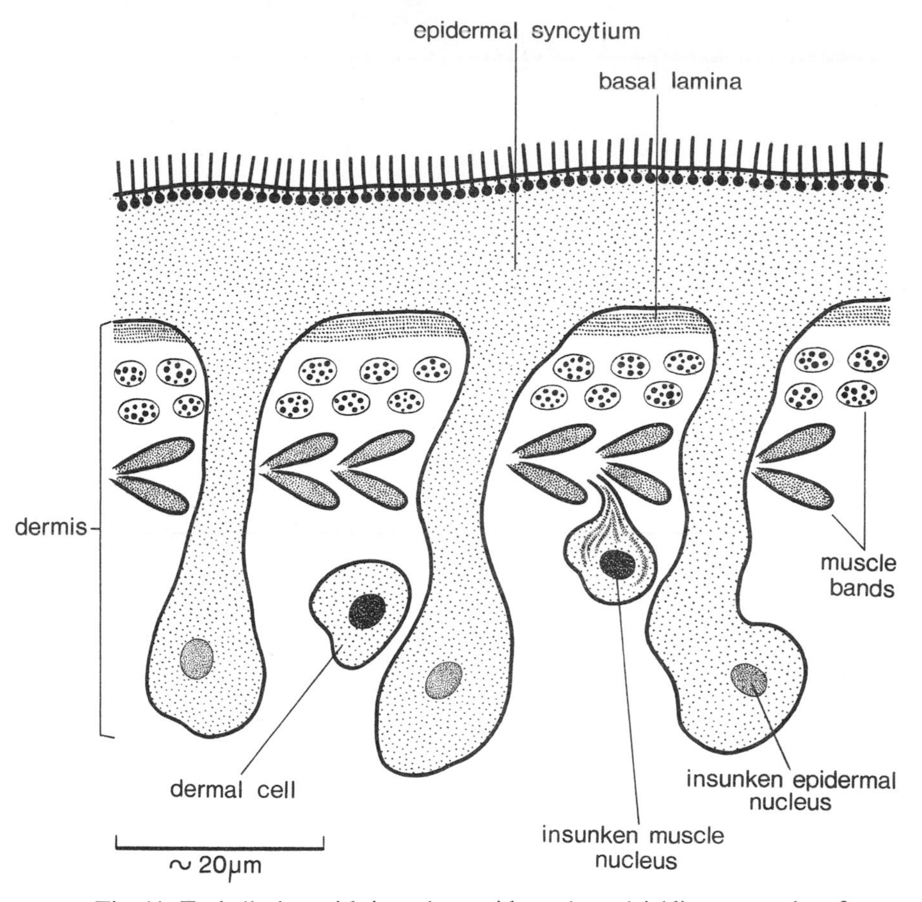

Fig. 11. Turbellarian with insunken epidermal, nuclei (diagrammatic, after Beklemishev).

Balanoglossus, and in many molluscs. (3) A layer of cuboidal cells with parallel upper and lower surfaces as occurs in many invertebrates including some turbellarians and arthropods. (4) A stratified epithelium characteristic of vertebrates, but which is found occasionally in invertebrates, as in the collarette of the Chaetognatha (arrow worms). Probably these epidermal types are of functional rather than evolutionary significance.

PARASITIC PLATYHELMINTHES

The histology of the integument in the parasitic flatworms, Cestoda and Trematoda, has been completely reappraised in the last decade as a result of examination by electron microscopy, and older accounts give wrong descriptions of the skin for both tapeworms and flukes.

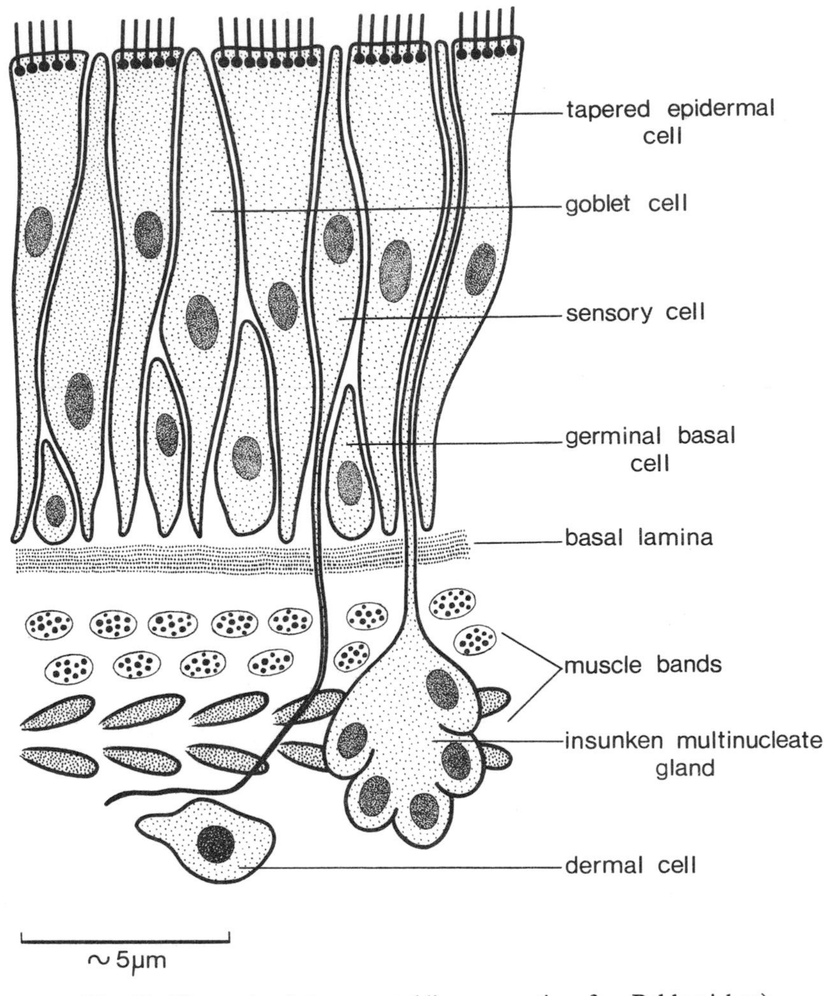

Fig. 12. Nemertine integument (diagrammatic, after Beklemishev).

By light microscopy, the diffuse staining reaction of the most superficial skin layer with complete absence of cell outlines or nuclei originally suggested that it was a cuticle which seemed illogically to rest directly on the dermis. The finding of mitochondria in this superficial layer at once showed that it was not a cuticle but part of an epidermal syncytium. Further investigation showed that the epidermal nuclei were situated much deeper in the dermis, connected with the surface syncytium by stalks of cytoplasm (Fig. 13; Plate 1). Essentially the organisation in these endoparasites is therefore the same as in many free-living flatworms, but there are no cilia and instead the outer surface is lined by absorptive microvilli

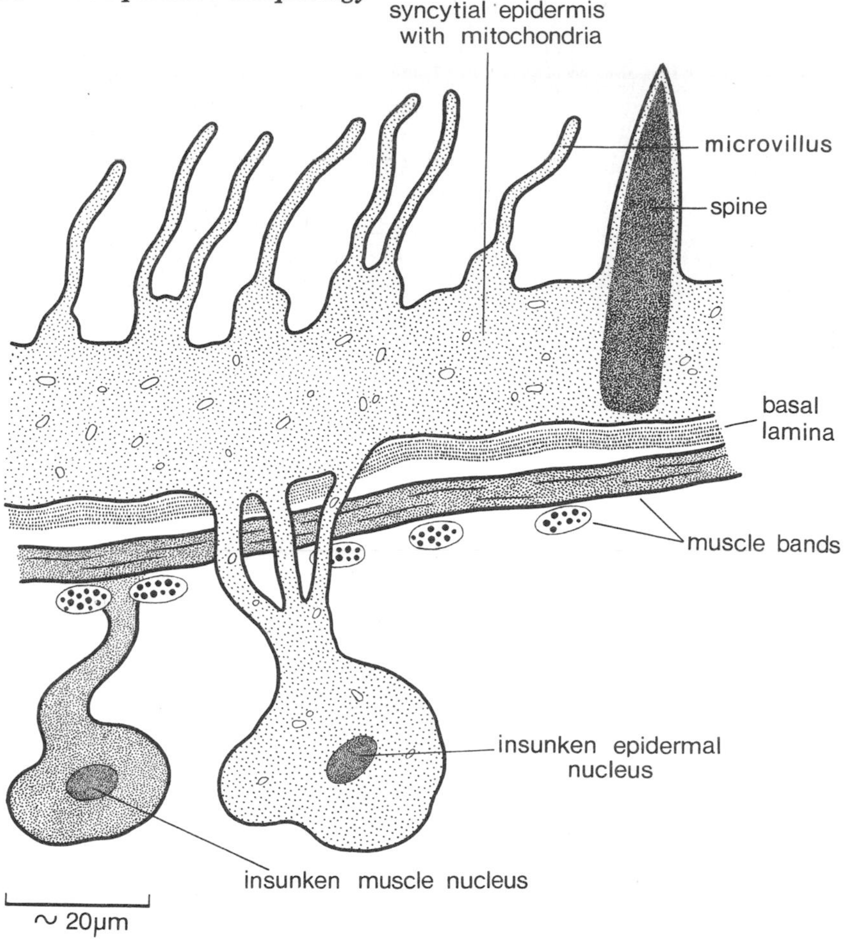

Fig. 13. Integument with insunken epidermal and muscle nuclei beneath the basal lamina; taken from various parasitic platyhelminthes.

as in the vertebrate intestinal mucosa. To emphasise its functional and structural peculiarities, the epidermal syncytial layer in both Cestoda and Trematoda is termed the tegument.

The skin in numerous species of tapeworms and flukes has been critically studied in the last few years, and both Cestoda and Trematoda have been shown to have a similar tegumental structure (Lee 1966). The epidermal syncytium contains a granular endoplasmic reticulum and is rich in both acid and alkaline phosphatase, the latter probably involved in transport across cell membranes. The appearance of microvacuoles just beneath the microvilli is indicative of pinocytosis of food particles. Under experimental conditions, labelled amino acids are rapidly absorbed through the

tegument against a concentration gradient, by active transport. Clearly, in tapeworms and flukes it functions as an inverted gut (Smyth 1969).

Although a cuticle is not developed in these parasitic worms, tough hard structures are sometimes formed within the tegument. Examples are the hooks on the adult scolex and on embryophores of tapeworms and the spines of flukes. The presence of cystine together with the general amino acid composition of these structures has suggested that keratin may be present. In vertebrates this durable protein is always formed as an intracellular product and it is possible that a form of keratin is produced by these parasites.

NEMERTINA

These are elongated acoelomate free-living marine worms which move by muscular locomotion. Ciliated cells and goblet cells make up the epidermis, and secretory cells grouped together as compound glands in the dermis with ducts to the surface are also present (Fig. 12). An interstitial syncytium-like network occurs around the basal processes of the tapered epidermal cells, but the nature of this remains to be determined (Hyman 1951*a*). Rhabdites occur in the epidermal cells of some species as in free-living platyhelminthes.

NEMATODA

This large group comprises the free-living and parasitic roundworms. Nematodes do not differ much in the broad micro-anatomy of the integument, but closer observation shows considerable variation in the basic organisation of the cuticle in different species (Bird 1971).

Generally, there is a thin dermal connective tissue layer over the deep musculature. External to this is a simple epidermis of cuboidal cells, often fused in a syncytium in adult worms. Cilia never occur anywhere in nematodes. On the outside is a prominent cuticle which is always much thicker than the epidermis which secretes it. In contrast to tapeworms and flukes, food is not absorbed through the nematode integument, and this permits a tough cuticle waterproofed with lipids.

The cuticle is mainly constructed of collagen which, as in dermal connective tissue, is associated with the acid mucopolysaccharides, hyaluronic acid and chondroitin sulphate. Chitin is absent but lipids occur. Basically the cuticle is divisible into four layers of varying relative thicknesses in different species (Lee 1966). The collagenous layers consist of a fibre region over the epidermis; above this is a matrix region and then an inner cortical layer. To the outside is an outer cortical layer of quinone-bonded sclerotin and a lipoidal epicuticle (Fig. 14).

The ultrastructure of the cuticle has been examined in a number of genera. In the common roundworm *Ascaris lumbricoides*, an intestinal

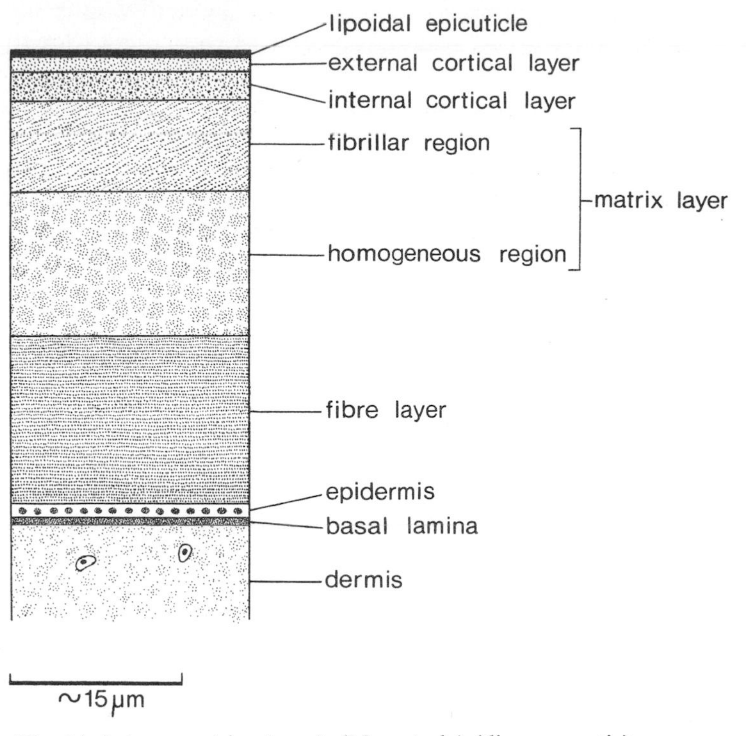

Fig. 14. Integument in *Ascaris* (Nematoda) (diagrammatic).

parasite of the pig and man, the outer cortical layer appears amorphous in electron micrographs and it is non-birefringent under the light microscope. It is constructed of quinone-bonded proteins which contain cystine and cysteine but no hydroxyproline, so collagen is absent although present deeper in the cuticle. Nor is it likely to contain keratin despite the cystine present because of the different X-ray pattern and its exocellular nature. It is best defined as a cystine-rich sclerotin until more is known about it. The sulphur-rich outer conchiolin layer in mollusc shells has similar properties.

The inner part of the cortex has a network of birefringent crisscrossing collagen fibres. The characteristic rings in *Ascaris* are due to grooves which extend no further than into the cortical region and allow the cuticle to be flexed. The matrix region is amorphous but contains some collagen, and the basal region shows crisscrossing collagen fibres.

A few nematodes examined have a markedly different cuticular ultrastructure from *Ascaris*. Thus, *Nippostrongylus*, an endoparasite of the rat, has a fluid matrix layer containing haemoglobin, and large fibres of collagen link the cortical and basal regions. In *Mermis*, whose larvae are para-

sitic in insects and with free-living adults, there is a thin amorphous cortex and a thick matrix layer containing two concentric layers of collagen fibres arranged obliquely so that they spiral around the length of the worm, one to the left and one to the right. The lower part of the matrix contains a network of fine crisscrossing fibres (Plate 2). Between the matrix collagen fibres is vacuolar non-fibrous material. The thin basal region is non-fibrous and similar to the vacuolar material of the matrix (Lee 1970).

Variation in the cuticle in different genera, therefore, largely consists of differences in distribution and arrangement of collagen fibres, probably determined by mechanical requirements in worms of different sizes. The tough nematode cuticle, apart from forming a protective layer, prevents the bulging of the worm during muscular contraction when the hydrostatic pressure of the body fluid is greatly increased.

The cuticular collagen in *Ascaris* contains less hydroxyproline than in vertebrate collagen. It has a characteristic collagen X-ray pattern, but transverse banding of the collagen fibres is poorly defined compared with vertebrate collagen. Cuticular collagen is digested by collagenase, but not the sulphur-rich sclerotin.

The presence of vertical pore canals in the nematode cuticle is controversial in *Ascaris*, but true canals are present in the cuticle of *Mermis* and in some other nematodes (Lee 1970; Bird 1971). Sometimes the pore penetrates from the epidermis to the cortex, suggestive of a glandular duct or a cell process. Species in which only the cortical ducts occur may have the lower parts of the canals obliterated during cuticle growth.

GROWTH AND MOULT OF NEMATODE CUTICLE

The nematode cuticle undergoes four larval moults to allow for body growth and no further syntheses of cuticle occur in the adult. Ecdysis appears to be mediated by neuro-hormones as in arthropods, but little is known about the endocrine control mechanisms in nematodes (Bird 1971).

In *Nippostrongylus*, the larval epidermis separates from the old cuticle prior to ecdysis and a new cuticle is then secreted beneath it (Lee 1970). The active epidermal cells have a prominent granular endoplasmic reticulum as in dermal fibroblasts which also exosecrete tropocollagen particles, the building blocks of collagen. The degree of fibre aggregation after this has left the cells is low in amorphous sites and greater in fibrous sites. As in connective tissue, this is dependent on the amount of tropocollagen secreted as well as on physical factors such as salt concentration, reaction with acid mucopolysaccharides, temperature, pH and lines of stress. In the plant parasite *Meloidogyne*, the lower layers of the old cuticle have been shown to be reabsorbed prior to moult, as in arthropods, probably after digestion in a protease-rich moulting fluid. Indeed, an

increase in leucine aminopeptidase in the epidermis, which drops at ecdysis, has been demonstrated in some species and is possibly involved in proteolysis.

Various hydrolytic enzymes occur in nematode cuticle and diffuse out from the epidermal cells during exosecretion.

GASTROTRICHA

This is a minor group of small marine acoelomate worms with a syncytial epidermis covered with a thin cuticle and with projecting spines. Ciliary tracts are used for locomotion.

KINORHYNCHA

These are similar to Gastrotricha and have a syncytial epidermis with a cuticle and spines, but no cilia.

NEMATOMORPHA (GORDID WORMS)

These have a degenerate gut. Larvae are parasitic in aquatic insects, and adult worms live in fresh water. A prominent cuticle is present (see Hyman 1951*b*).

ROTIFERA

These are small fresh-water animals. They have a syncytial epidermis except for locomotory ciliated cells which have septate desmosomes. A hard cuticle (the lorica) is present over most of the body (Koehler 1965). This contains beta chitin, acid mucopolysaccharides and sclerotin. The cuticle has a complex structure and is completed shortly before birth. Adults do not secrete a new cuticle. The surface of the epidermal syncytium adjacent to the cuticle shows peculiar flask-shaped depressions of undetermined function.

ACANTHOCEPHALA (THORNY-HEADED WORMS)

These small unsegmented worms are all parasitic in aquatic animals. There is no gut and food is absorbed through the integument. The cuticle contains acid mucopolysaccharides, lipids and sulphur amino acids, but not collagen. The characteristic proboscis bears hooks of different composition (Florkin and Scheer 1969*a*). The epidermis is unique in that the syncytium is pierced by a network of canals lined by cuticle which communicate directly with the surface for food absorption.

PRIAPULIDA

These are small unsegmented free-living marine animals with a simple cellular epidermis which has an abundant granular endoplasmic reticulum. There is a thick fibrous cuticle of unknown composition. The ventral nerve cords are embedded between the tapering ends of the epidermal cells. The dermal side of the epidermis is pitted with pores which lead to a labyrinth of intercellular spaces of unknown function (Moritz and Storch 1970).

ENTOPROCTA

These are small colonial sedentary polyp-like acoelomate animals once grouped with the Ectoprocta as the Polyzoa. Entoprocta have a simple cuboidal cellular epidermis covered by a cuticle of undetermined composition. The cuticle is thickened over the stolons and in the calyx shields of each individual. Glandular cells connect with hollow cuticular spines. The inner surface of the tentacles does not develop a cuticle and is ciliated for the transport of food to the mouth (Atkins 1932). Sensory receptor cells are particularly numerous in the tentacular epidermis (Hyman 1951*b*).

5

THE INTEGUMENT OF COELOMATE INVERTEBRATES

ECTOPROCTA

Ectoprocta are sedentary polyp-like aquatic animals, with free-swimming ciliated trochosphere larvae. Each individual has a circumoral ring of ciliated tentacles, the lophophore, and the sides have a thick cuticle, the zoecium, brittle in encrusting species through deposition of calcite. Hyman (1958) found chitin in the superficial cuticle while calcification was mainly deeper. Colonies are produced by the packing together of neighbouring zoecia with no organic connection between individuals as in Entoprocta and colonial hydrozoa.

BRACHIOPODA

These marine animals evolved a mineralised shell independently of molluscs. They have ciliated trochosphere larvae. As in bivalve molluscs, the two parts of the shell are joined by a calcified flexible hinge. However, while in brachiopods the valves are formed over the dorsal and ventral surfaces, in molluscs they are bilateral. The soft part of the integument is referred to as the mantle. The mantle epidermis, including the lophophore, is ciliated and can be completely enclosed by the shut valves. Planktonic food is sucked into the mantle cavity by cilia beating towards the mouth, and water and waste are removed by cilia beating in the opposite direction. Chitinous spines sometimes occur around the mantle edge and there is also a muscular foot covered with a thick chitinous cuticle.

The shell is a calcified epidermal exosecretion which begins to be laid down in the late larval stage when the animal is only a millimetre or so in length and continues throughout the life of the individual, which may eventually reach several centimetres. The complex relationships between growth and form in brachiopod and mollusc shells with their great increase in size over the years is discussed by Thompson (1942). As the brachiopod grows, the shell-forming epidermis increases its surface area, probably by cell division around the margin of the valves. New shell is formed in this region and deposition also continues beneath the valves, which in consequence get thicker each year (Fig. 15).

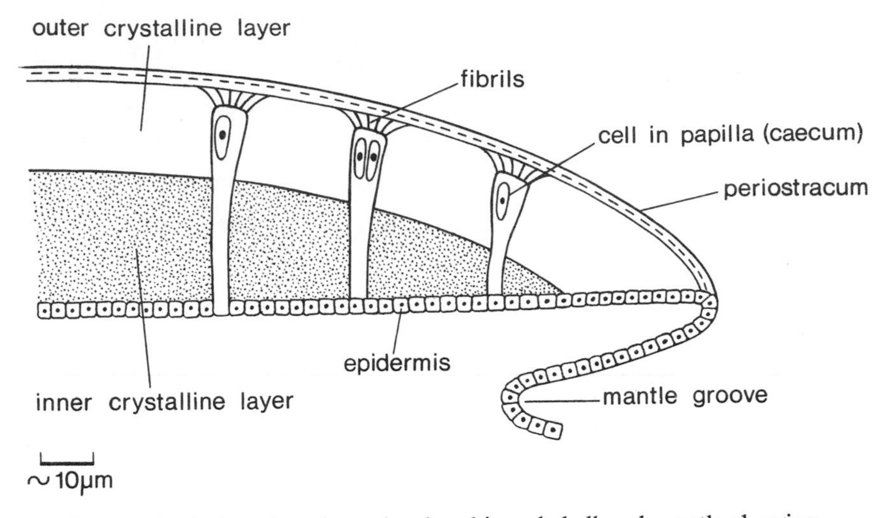

Fig. 15. Vertical section through a brachiopod shell and mantle showing crystal columns over formative epidermal cells.

The first stage of shell formation is the exocellular secretion of a thin cuticle termed the periostracum, composed of a non-orientated sclero-protein linked to beta chitin. Thenceforth, periostracum formation is confined to the mantle edge. A thin outer mineralised layer of vertically orientated calcite or calcium phosphate crystals is then laid down at the shell edge under the previously formed periostracum. As the shell gets thicker, canals containing cells continue to connect the epidermis with the periostracum in many brachiopods. Whether or not these canals become fibrous is uncertain; if so, they may help to anchor the periostracum or they might be sensory processes. The presence or absence of canals is used in classification. Thus, in some species they remain throughout the depth of the adult shell (perforate forms). In others, the perforations do not extend into the inner calcified layer, possibly because the canals become filled in (semi-perforate forms). In others, they never occur at any stage (imperforate forms).

Under the older part of the shell, a thick inner calcified layer is laid down beneath the outer crystalline layer. This has a crystal orientation oblique to the mantle surface and in terebratulids it forms a skeletal framework for the lophophore (Williams 1956).

Shell mineralisation probably occurs on an organic matrix of secreted protein particles, and the arrangement of these proteins may determine the crystal orientation. Calcium, and in some species also phosphate, are concentrated by the animal from the sea water and, together with carbonate ions from respiration, pass out of the epidermis in solution and

crystallise out to form calcium carbonate or phosphate. In the most primitive genus *Lingula*, the valves have a chitinous layer in place of the outer calcareous layer (Hyman 1959).

The shells of brachiopods and molluscs, although both constructed of three layers, show differences in organisation and composition.

CHAETOGNATHA (ARROW WORMS)

These small, rigid, rod-shaped marine animals are represented by a single genus *Sagitta*, with pelagic marine species. Hadzi (1963) and Hyman (1940) place them close to the Brachiopoda on the phylogenetic path to Chordata.

Cilia occur only in the planktonic larvae. The whole epidermis in the adult is covered with a thin non-chitinous cuticle. The epidermis is a single layer of cuboidal cells, except over the head and collarette behind the head where it is stratified and several cells in depth. A stratified epidermis is of rare occurrence in invertebrates. Distinctive columnar epidermal cells exosecrete the hard jaws and circumoral spines which contain beta chitin (Hyman 1958). Lateral fins are made up of a dermal collagenous plate covered on either side by epidermis and cuticle. Glandular goblet cells occur in the collarette (Hyman 1959).

PHORONIDEA

This is a small group of solitary and sedentary marine animals which catch planktonic food by means of a ciliated lophophore. The remainder of the epidermis secretes a thin cuticle and also contains glandular cells. There is a ciliated trochosphere-type larva.

ECHIURIDA

These are believed to be related to the Annelida but show few signs of metameric segmentation. There is a ciliated trochosphere larva. All are marine and sexually dimorphic with sac-like sedentary females. Adult males in contrast are small ciliated turbellarian-like animals which occur on the skin surface of the very much larger females. The latter have a single pre-oral tentacle-like prostomium. There is a single pair of anterior setae and a few posterior setae. The epidermis in the female of *Bonellia* is a single layer of cuboidal cells which secretes a thin cuticle. Individual glandular cells occur in the epidermis and multicellular glands with ducts to the skin surface are found in the dermis.

SIPUNCULOIDEA

These appear related to annelid worms, and they have a ciliated trochosphere larva. There is a sac-like body in the sexually dimorphic females, as in *Bonellia*, but no prostomium. The epidermis is a single layer of cuboidal cells with a rough endoplasmic reticulum. The fibrous cuticle in its microstructure resembles that in annelids. The tentacles are ciliated (Mortiz and Storch 1970). The spines are thicker and longer than the setae of Echiurida.

ANNELIDA

The true segmented worms are divided into three classes: the Polychaeta, marine species with parapodia bearing prominent setae, the Oligochaeta which inhabit fresh water or are terrestrial earthworms with small setae and no parapodia, and the Hirudinea, leeches mostly devoid of setae.

The epidermis is a single layer of cuboidal cells connected by septate desmosomes. In the earthworm *Lumbricus* it is made up of cuticle-secreting cells and mucous goblet cells. Small germinal replacement cells sometimes form a discontinuous lower layer. The cuticular surface of the epidermis in the earthworm has a prominent border of microvilli. The thick cuticle completely covers the body (Dales 1967). Ciliated tracts occur in aquatic species (Plate 3).

The annelid cuticle is composed of collagen with an acid mucopolysaccharide matrix, as in nematodes (Watson 1958). Cuticular collagen fibres are arranged in layers with fibres of alternate layers at right angles to each other (Coggeshall 1966). Above the collagen layers is a thin amorphous epicuticle composed of quinone-bonded protein. The whole cuticle is devoid of chitin, but sclerotin in the setae is linked to beta chitin, and these are exosecreted by morphologically distinct epidermal cells. In the polychaete *Aphrodite*, 35 per cent of the dry weight of setae is chitin. In ultrastructure, each seta is constructed of several parallel microtubules, the walls of which are fused towards the tip (Orrhage 1971; Florkin and Scheer 1969*b*).

Storch and Welsch (1970) showed by electron microscopy in a variety of polychaetes that the epidermal cells have a rough endoplasmic reticulum with bound ribosomes as well as free ribosomes. Bundles of tonofilaments are present and peculiar vacuoles (or microtubules) frequently occur in the cytoplasm, which in some species are of sufficient size to displace the nucleus. Their significance is not known. Fibrillar cuticular material often found in the crypts of the microvilli is absent in a number of the species which secrete a covering tube (Serpulimorpha). In terebellids, also tube-dwelling polychaetes, the epidermis secretes an organic cement substance which sets and binds together mineral particles such as sand grains

obtained from the substratum. In serpulids, calcite is laid down in the cement around the epidermis, but in *Chaetopterus* the cement sets to form a parchment-like membrane and is non-mineralised. Clearly the polychaete tube is cuticular in nature, although sometimes strengthened by material from the substratum. In many tube dwellers, a median ciliated groove runs the length of the worm for removal of waste. The circumoral tentacles in sedentary tube dwellers are also ciliated for gathering planktonic food.

In annelids there is a clearly distinct dermis demarcated below the body musculature. In oligochaetes relatively few dermal cells occur. Typical fibrocytes occur in the leech dermis together with particularly prominent melanophores. The dermis in leeches contains a prominent network of storage blood capillaries, some of which loop up between the epidermal cells towards the cuticle (Mann 1962).

Sensory receptor cells often grouped together as sensillae occur in the epidermis of many annelids.

POGONOPHORA

This small marine phylum contains sedentary worm-like animals which form chitinous tubes and have circumoral tentacles, as in tube-dwelling polychaetes. The epidermis also has microvilli with fibrous material deposited in the crypts (Ivanov 1963; Brunet and Carlisle 1958). Nevertheless, despite a superficial resemblance to annelids, the Pogonophora on other grounds are placed away from them near the Lophophora on the line to chordate evolution (Hadzi 1963).

ECHINODERMATA

The most characteristic feature of the echinoderm integument is the calcified skeleton, termed the stereom, laid down in the dermal collagenous connective tissue and made up of a network of rigid calcareous plates enclosing an interconnecting fluid-filled space. The stereom is made up of 71–95 per cent calcium carbonate and 3–15 per cent magnesium carbonate (Florkin and Scheer 1969*b*). In sea urchins (Echinoidea) it forms projecting spines covered by a thin layer of collagenous dermis and epidermis. Pilkington (1969) showed that in sea urchins, dermal amoeboid cells (sclerocytes) endosecrete and repair the calcite skeleton. These peculiar lipid-rich cells fuse to form a giant syncytium which contains the stereom (Fig. 16). Crystallisation occurs when calcium and carbonate ions are received from the extracellular stereom fluid. Initially, spicules are separate but later coalesce to form plates.

The epidermis in echinoderms is either a single layer of columnar cells

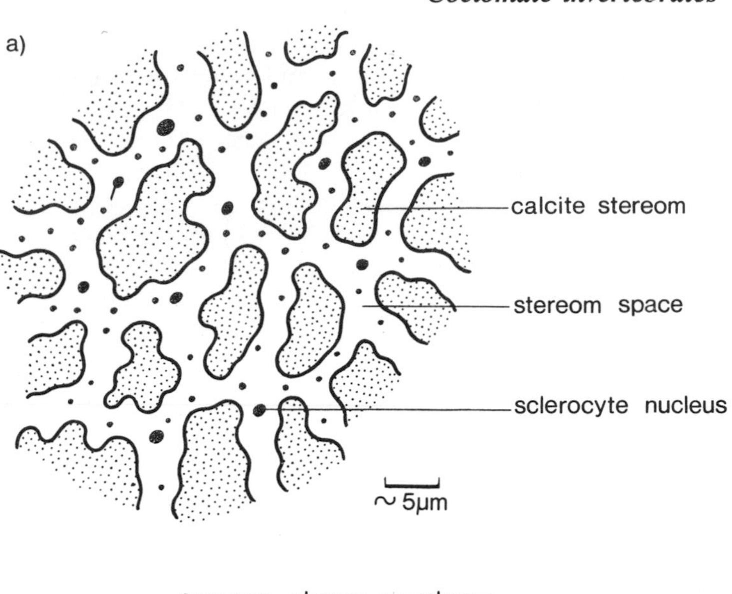

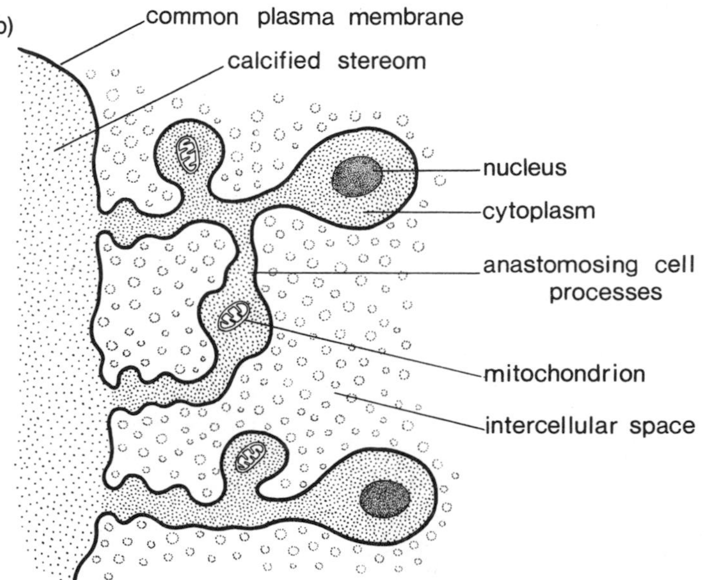

Fig. 16. Low and high magnifications of the echinoderm stereom showing apparent crystallisation within the sclerocyte syncytium (diagrammatic, after Pilkington).

held together by septate desmosomes or a syncytium. In starfish (Asteroidea) a two-layered cuticle is formed, and cuticle-secreting cells in the epidermis are interspersed with mucous goblet cells. Compound glands also occur in the dermis with ducts to the surface. Only the ambulacral grooves have cilia. De Souza Santos and da Silva Sasso (1970) found that in the tube feet the cuticle, which contains both acid and neutral mucopolysaccharides, is formed by an epidermal syncytium with nuclei suspended on stalks in the dermis as in some platyhelminthes. The syncytium bears secretory microvilli.

In brittlestars (Ophiuroidea) there is a thin cuticle over an epidermal syncytium, but the skin is less glandular than in Asteroidea. In sea cucumbers (Holothuroidea), the epidermis is composed of columnar cells interspersed with glandular goblet cells. There is a well-defined cuticle and cilia are absent. Echinoidea in contrast have a completely ciliated epidermis except over the pedal discs which have numerous mucous cells. Sea lilies (Crinoidea) mostly have a syncytial epidermis and a cuticle, with cilia confined to the ambulacral grooves (Hyman 1955).

ARTHROPODA

The extensive literature on the arthropod integument was comprehensively reviewed by Richards (1951), and later work has been the subject of several review articles (see Chapman 1969). Function as a respiratory surface is confined to specialised gills absent in insects, and it is never ciliated.

The epidermis consists of a single layer of cuboidal cells with other scattered glandular cells which secrete a thick tough cuticle made up of fibrous sclerotin stabilised by quinone and other phenolic cross-linkages, and strengthened by alpha chitin, distinct from the beta form of lower invertebrates (Rudall 1963). Collagen never occurs in the cuticle, in contrast to annelids. In many arthropods the cuticle is further strengthened by deposition of calcite; this is important for its skeletal function. A characteristic feature of arthropods is the endophragmal skeleton formed by deep invaginations of the epidermis at the joints. In these sites rigid cuticle in the form of apodemes secreted between opposing epidermal surfaces is used for muscle attachments. A disadvantage of an exoskeleton is that it will not permit growth of the animal and it has to be moulted and replaced as the body increases in size. Just after ecdysis the soft newly-formed cuticle is for a time unable to provide firm attachments for muscles, and arthropods are particularly vulnerable at this stage since they are temporarily immobilised. This remains a problem throughout the life of crustaceans, but in arachnids and insects it is overcome by restriction of growth and ecdysis to the larval stages.

INSECTA

Epidermal exosecretion

The epidermis, in addition to cuboidal cells, contains scattered larger oenocytes. The cuboidal cells are thought to secrete cuticular proteins and the oenocytes cuticular lipids, waxes and possibly chitin (Wigglesworth 1970). Pores extend vertically through the cuticle and in development may contain cytoplasmic processes of epidermal cells which dry up to leave canals. These are utilised for transport to the outer part of the cuticle and epicuticle of waxes and phenolic substances involved in protein tanning. Unicellular glands secrete the surface cement, and multicellular glands with ducts to the surface also occur in the dermis. The secretory products include waxes, scents, toxic repellents, and possibly proteolytic enzymes of moulting fluid which break down the old endocuticle prior to ecdysis.

The insect cuticle

This is a complex layered structure (Fogal and Fraenkel 1970; Rockstein 1964; Wigglesworth 1965) (Fig. 17).

Epicuticle. On the outside over the rigid cuticle is a thin epicuticle 1–4 μm in depth. This contains hydrophobic lipids and aliphatic waxes but not chitin, and constitutes the barrier to diffusion of water into and out of the insect. On the outer surface is a layer of cement and then a monolayer of wax molecules, possibly orientated so that hydrophobic aliphatic groups face outwards and hydrophilic groups inwards towards an orientated layer of lipoprotein, termed cuticulin. Beneath this is a thicker homogeneous protein layer (Plate 4).

Rigid cuticular regions. These may be 200 μm thick and contain alternate layers of chitin and fibroin proteins bonded together as in plywood. The new cuticle is subdivisible into a hard, poorly stainable exocuticle in which the fibroins are tanned to form sclerotins, a similar but softer mesocuticle, later incorporated in exocuticle, and a still softer, readily stainable, untanned endocuticle. The latter has a characteristic lamellar microscopic structure due to the ordered arrangement of chitin filaments.

Joint cuticle. These flexible areas contain the elastic protein resilin (Andersen and Weis-Fogh 1964).

Cuticle synthesis and ecdysis. The epidermis in adult insects is physiologically inactive since cuticle synthesis has ceased apart from renewal of wax and lipids. Cuticle formation must therefore be studied in the larval integument or in adults in which the cuticle is repaired after damage.

Before each moult the epidermal cells undergo rapid lateral division. As a result, they appear columnar, but after ecdysis lateral expansion occurs and cells resume a cuboidal form (Smith 1968). The separation of the old cuticle from the epidermis (apolysis) is initiated by changes in ten-

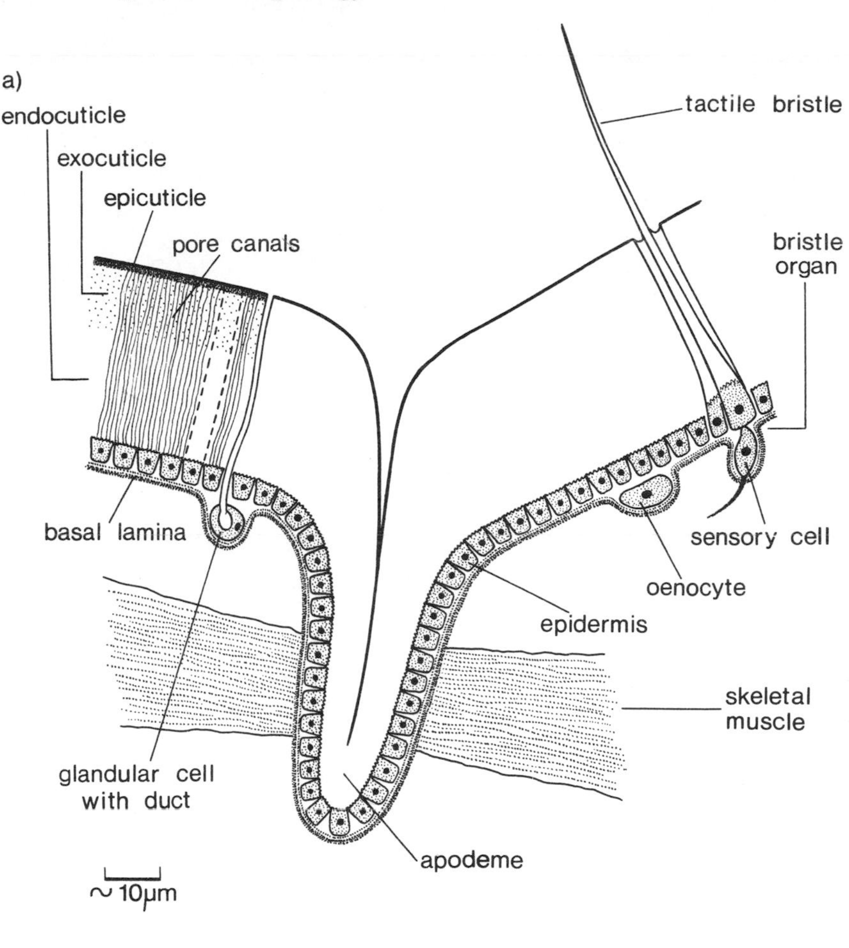

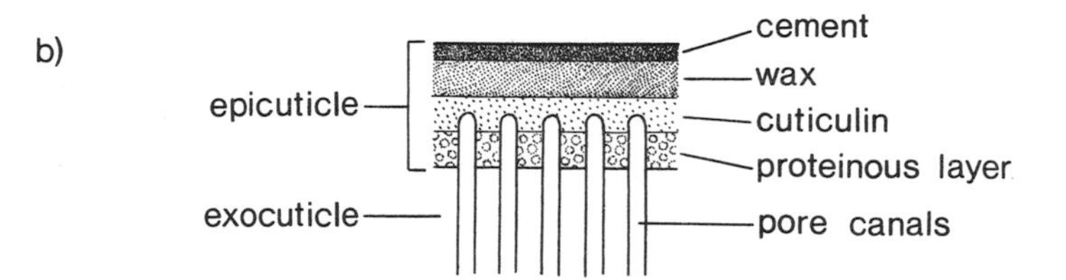

Fig. 17. (*a*) Insect integument showing structures from various sites. Cuticle formed by one epidermal cell between dotted lines. (*b*) Higher magnification of insect cuticular surface layers showing probable arrangement (after Wigglesworth).

sion along the junction resulting from epidermal lateral growth. Moulting fluid is secreted which contains inactive precursors of protease and chitinase enzymes, and forms a layer between the epidermis and old cuticle. Meanwhile, the epidermal cells develop a prominent rough endoplasmic reticulum with bound ribosomes and they secrete new cuticular waxes, proteins and chitin. The resistant epicuticle is first laid down which protects the epidermal cells beneath from the moulting fluid. The old endocuticle which lacks resistant phenolic linkages is then broken down to soluble constituents mostly reabsorbed by the larva. In this way, some 90 per cent of the old cuticle is conserved (Chapman 1969; Noble-Nesbitt 1963). After epicuticle formation, unbonded procuticle is laid down. The outer region of the procuticle later becomes tanned and constitutes the new exocuticle and mesocuticle, but the inner part remains unbonded as the endocuticle. The old epicuticle and exocuticle now split along weak ecdysal lines and are shed. This process is facilitated by the larva swallowing air.

During cuticle formation the epidermal cells develop microvilli and filamentous processes and these come to lie embedded in the cuticle. Once completed, the microvilli disappear. Perhaps they form the pore canals.

Tanning of the new cuticle. This is at first still pliable. Hardening and stabilisation of the protein chain molecules in insects, as in many other invertebrates, is due to phenolic tanning to give sclerotins. Sulphur amino acids do not occur and calcification in insect cuticle is negligible. Shortly before moult, there is an increase in the level of blood tyrosine. Oxidation by tyrosinase of tyrosine to Dopa followed by decarboxylation and acetylation to *N*-acetyldopamine are also controlled by specific enzymes. *N*-acetyldopamine diffuses out through the newly formed cuticle, possibly by way of the pore canals to its outer layers. Cross-linkages between protein chains are thought to occur when the quinone polymer links with amino side groups, and polypeptides are probably bonded together end to end through links between quinone and *N*-terminal amino acids. Darkening of the hardened cuticle occurs in the process which is related to melanin formation. It is not known which cells produce free cuticular phenol oxidases, except for the cockroach ootheca where the left collateral gland is the source. Several different fibroin proteins occur in the cuticle. Other derivatives of *N*-acetyldopamine which form cross-linkages are catechols which become linked by their aliphatic side-chains to protein. The chemistry and hormonal control of the tanning process are discussed elsewhere. (See chapter 15.)

Injected labelled *N*-acetyldopamine with H^3 in the acetyl group is rapidly taken up into the cuticle during growth, and labelled injected substances combined with examination of cuticular hydrolysates have been widely used in studies of cuticle synthesis.

Chitinisation. The nitrogenous mucopolysaccharide alpha chitin represents 25–50 per cent of the dry weight of insect cuticle, and gives additional tensile strength (Rudall 1963).

Sensory organelles of insects

Various integumental sensillae have been evolved by insects to overcome the problem of a rigid exoskeleton. Trichoid sensillae (bristle organs) are hair-like projections of cuticular material formed by epidermal trichogen cells. These act as mechanoreceptors and chemoreceptors. In some other types of sensillae the overlying cuticle remains thin, and nerve axons pass up between the epidermal cells to end just beneath it. The antenna integument is particularly well supplied with sensillae.

CRUSTACEA

Morphology

Although in the Decapoda, such as crabs, lobsters, prawns and shrimps, the thick cuticle has become a brittle carapace strongly mineralised with calcite, this is not typical of Crustacea as a whole, and in the much smaller Branchiopoda such as *Daphnia*, it is only weakly calcified. In these species strength and support for muscles is achieved by sclerotisation. As in insects, sulphur amino acids do not occur to any extent in crustacean cuticle.

The thick decapod cuticle is a layered structure which, apart from its greater mineralisation, resembles insect cuticle. Externally there is a thin non-chitinous, quinone-tanned epicuticle which contains lipids and has a surface layer of aliphatic wax molecules, although not so well orientated as in insects. The strength of the epicuticle was not greatly impaired by decalcification in acid (Dennell 1960) and its toughness is due to sclerotisation. As in insects, the epicuticle acts as a barrier to water penetration (Lockwood 1968).

Beneath the epicuticle is a dark brown pigmented layer containing free melanin in addition to quinone-tanned protein. In surface view it shows an hexagonal pattern of columns of calcite crystals. These are continuous through different layers and each is formed over an hexagonal epidermal cell. Chitin and sclerotin in this region are present mainly in the septae between these columns. Beneath the pigmented layer is a less melanotic, calcified layer rich in chitin and sclerotin which forms the greater thickness of the carapace. The combined pigmented and less pigmented calcified layers are homologous with the insect exocuticle (Dennell 1960).

The thin endocuticle in decapods is uncalcified even in the hexagonal columns and, as in insects, the protein is untanned. Calcite probably crystallises out on some protein secreted by the hexagonal cells. A lamellar

arrangement of chitin–protein filaments occurs in the decapod endocuticle, as in the insects. The dry weight of decapod cuticle is 60–80 per cent chitin (Dennell 1960), which is slightly greater than in insect cuticle (Wigglesworth 1965). Flexible joints are uncalcified and contain resilin.

Moult and cuticle formation in Decapoda. Growth and moult continue well into adult life in Crustacea. The length of the moult and intervals between moults vary with species, age and environmental temperature. The prawn *Leander* moults every 14–18 days, and moults occur every two days in the ocean barnacle *Balanus*. Travis (1955) has examined the moult cycle in decapods. The stages are as follows:

1. Immediate post-moult period. The epicuticle and exocuticle are completed. Calcification of the new exocuticle begins, but it is too soft for muscle support. No feeding. Water absorption through cuticle is increased.
2. Sclerotisation and calcification continued. No feeding.
3. Calcification of the new cuticle nearly completed. Water absorption is reduced. Feeding recommences.
4. New cuticle completed. Intermoult period. Low water absorption.
5. Preparation for moult. Calcium of old cuticle is reabsorbed, especially along dehiscence lines, and untanned protein of endocuticle is digested by moulting fluid enzymes. Phagocytic cells migrate into the endocuticle and remove particulate matter.
6. Epidermal mitosis and lateral growth occurs. The old cuticle is separated from the newly formed epicuticle by a layer of moulting fluid. Calcium salts are reabsorbed from the old cuticle and enzymes break down untanned protein and chitin. Phagocytes migrate into old endocuticle and remove particulate matter. Water absorption increased.
7. New pigmented and unpigmented calcified layers laid down.
8. Ecdysis of old cuticle occurs.

Cuticular-forming material may be stored in the hepatopancreas in decapods, which also shows cyclical changes.

ARACHNIDA

In contrast to insects and crustaceans which do not have cuticular cysteine, spiders, mites and king crabs have an additional outer hyaline layer beneath the epicuticle rich in this bound sulphydryl amino acid. This layer is absent in the other classes. Hughes (1959) did not find in hydrolysates of mite cuticle the disulphide-containing amino acid cystine which is present in keratin. He suggested that the tanning process of arachnids probably involves the formation of quinone–SH complexes.

Arachnid cuticles are only weakly calcified and in contrast to insects and

crustaceans they contain the more primitive beta form of chitin (Richards 1951). Aliphatic waxes occur in some species but resilin has not been detected. Cartilage occurs in the gill dermis of king crabs.

OTHER ARTHROPODS

Myriapoda are primitive arthropods with beta chitin. They do not have a waxy epicuticle but have numerous dermal glands which secrete lipids over the cuticle. Resilin is found in joints of centipedes. In the most primitive living group, the Onychophora represented by *Peripatus*, neither wax nor lipids occur in the cuticle. Otherwise, it is built on the basic arthropod plan with an epicuticle and beneath it a sclerotised cuticle. This contains beta chitin (Robson 1964). The most aberrant condition is seen in the small fresh-water Tardigrada which appear to have only a non-chitinous epicuticle which is frequently moulted.

ARTHROPOD SILK GLANDS

Large compound protein-secreting glands occur in many arthropods. These are epidermal derivatives which lie in the dermis, and connect with the skin surface where they discharge filaments of fibroin protein as in the silkmoth *Bombyx*, and in spiders. Collagen occurs in Hymenoptera silk and chitin-bound protein in the praying mantis. Silk production involves an exceptionally high rate of protein synthesis (Peakall 1969).

MOLLUSCA

PRIMITIVE GROUPS

Most molluscs are marine and there are relatively few fresh-water, and still fewer land, species.

In the primitive worm-like Aplacophora, the epidermis secretes a homogeneous non-chitinous cuticle in which calcite spicules are sometimes embedded. The Polyplacophora, represented by *Chiton*, have in some regions a thin chitinous cuticle containing calcite spicules, and over the dorsum a series of more strongly calcified valves: the first appearance of a shell (Hyman 1967). Each valve has a thin outer layer, the periostracum, which probably represents the primitive cuticle, and beneath it is a thick mineralised crystalline region. The phylogeny of the mollusc shell through thickening and calcification of the cuticle is demonstrated in the different classes of molluscs.

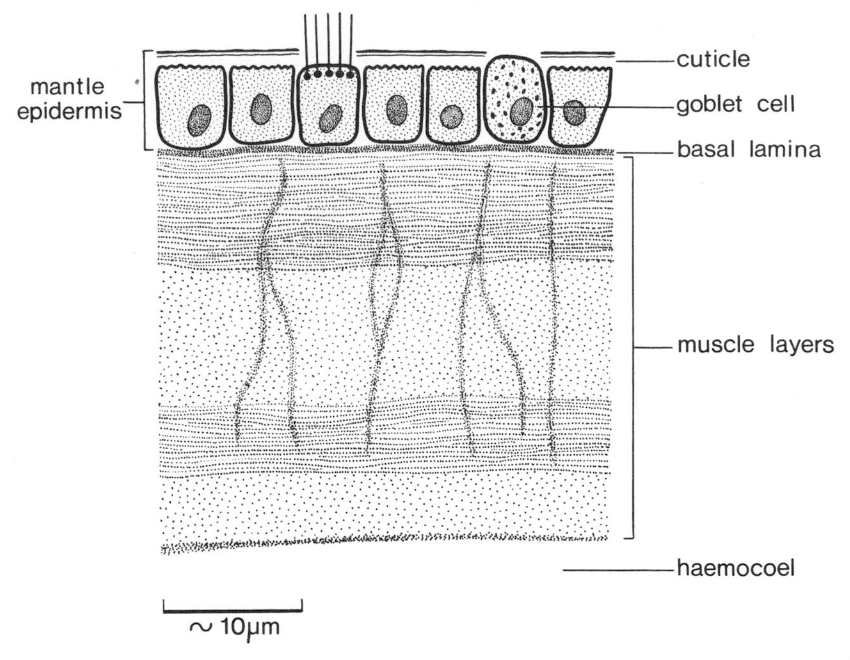

Fig. 18. Generalised mantle integument in a gastropod mollusc (after Fretter and Graham). The body wall musculature grades into dermal musculature.

HIGHER GROUPS

In both gastropods and lamellibranchs the soft epidermis is ciliated or bears a thin cuticle over microvilli and is termed the mantle (Fig. 18). It is a columnar layer and contains mucous goblet cells (Graham 1957).

Ciliary motion in filter-feeding lamellibranchs sucks in water and food particles towards the mouth and removes waste products. The mantle is also a respiratory surface and gills increase the surface area.

Compound mucous glands in the dermis and deeper tissue discharge to the surface in certain sites such as the foot in land snails and produce the characteristic silvery tracks. The wafer-like operculum which covers the shell entrance in hibernating snails is also a glandular secretion. An adaptation is acid mucus used to erode calcified rock by rock-boring species (Jaccarini, Bannister and Micallef 1968).

The thin dermis in molluscs is demarcated below by the body-wall musculature.

Physical protection is provided by the shell which is large enough in most molluscs to conceal the soft parts of the animal. The shell contains some protein but is largely built of calcium carbonate in its various crystalline polymorphic forms.

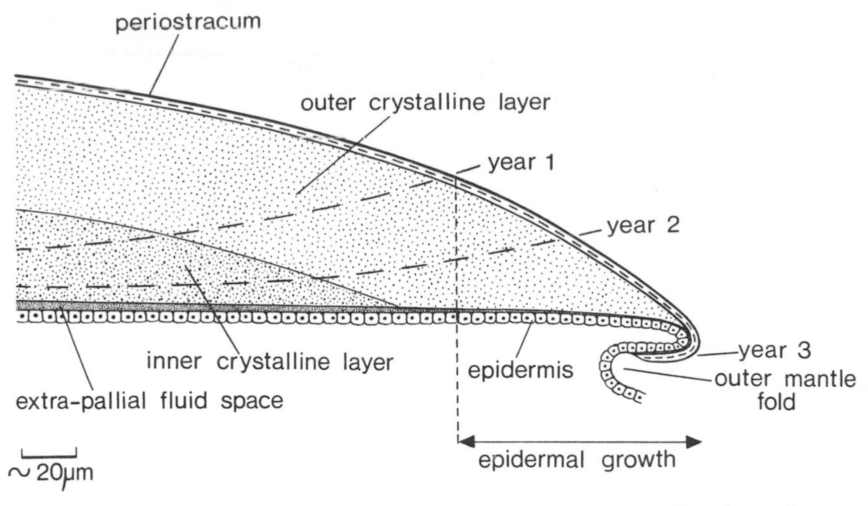

Fig. 19. Vertical section through lamellibranch mollusc shell and mantle.

Development

Aquatic species pass through ciliated trochosphere and veliger larval stages. The shell is laid down over the shell organ of the veliger, the first part formed being the periostracum. Later the calcified shell is formed beneath it and further growth takes place from the shell margin. Therefore, presumably epidermal mitosis necessary for increase in shell circumference is confined to the mantle groove which in consequence moves further and further away from the shell midpoint. In the life of a mollusc, the shell increases greatly in size, as in brachiopods.

The shell is built on the same basic pattern in both gastropods and lamellibranchs which comprise the majority of living molluscs. In bivalve species, the hinge made of weakly calcified sclerotin is always dorsal and the valves bilateral, in contrast to brachiopods.

The shell is made up of three layers in higher molluscs (Fig. 19). On the outer surface is a thin uncalcified periostracum, largely composed of various quinone-tanned sclerotins termed conchiolin, which occur in alpha and beta forms. Hare (1963) showed in the mussel *Mytilus edulis* that proteins with different amino acid compositions occur at different levels in the shell. Proline and cystine are confined to the periostracum and relatively fewer acidic than basic amino acids occur than in the mineralised shell. The sulphur-rich periostracum appears similar to the nematode outer cortical layer, but it is unlikely that keratin is present as an extra-cellular product, as has been suggested, and the different X-ray pattern excludes collagen (Wilbur 1964).

The mantle groove epidermis which secretes the adult periostracum has

a columnar epithelium rich in RNA, as shown in the fresh-water snail *Limnaea*. The mantle in this region contains oxidation products of tyrosine, and further oxidation to orthoquinones within the periostracum is followed by linkage with conchiolin protein (Brown 1950) as in arthropods. Various phenolic oxidases for sclerotisation occur in different molluscs (Timmermans 1969; Beedham 1958). The brown colour of many mollusc shells is produced by this tanning process.

Immediately beneath the periostracum is the thick outer calcified or prismatic layer. This is formed by the epidermis just within the shell edge and has similar RNA-rich cells to those of the mantle groove (Timmermans 1969). The prismatic layer contains much less protein than the periostracum, although possibly as much protein is originally formed but is diluted by the large amount of calcium carbonate which crystallises out. This region is the thickest part of the shell in contrast to brachiopods, and is generally made up of vertically-orientated crystals of calcite.

The inner calcified (nacreous) layer is laid down still further from the shell edge and usually contains crystals of calcium carbonate in the form of aragonite, deposited in horizontal lamellae with alternate mineral and protein layers. These lamellae are between 0.3 and 2.3 μm thick in marine lamellibranchs. When the mantle is lost in the dried shell, the nacreous layer has a pearly irridescence by reflected light and is called mother-of-pearl. Precious pearls are abnormal growths of nacreous layer induced by foci of epidermal irritation in the soft mantle, either of natural origin or purposely introduced as in cultured pearls. These occur in many lamellibranches besides the oriental pearl oyster. In this pathological process the mantle epidermis changes from a mucous epithelium to a shell-forming organ. In a few species the whole thickness of the mineralised shell is made up of aragonite with no calcite, and occasionally valerite, another polymorphic form of calcium carbonate, is present.

The epidermis over which the inner calcified layer is laid down has cuboidal or more often flattened cells and contains much less RNA than in the other two regions. Phosphatase enzymes frequently occur in the shell-forming epidermis and are probably concerned in transport across cell membranes (Timmermans 1969).

Mechanism of shell formation

Untanned matrix protein and mucopolysaccharides are secreted in solution by the epidermis to form the extra-pallial fluid which then separates the epidermis from the earlier-formed periostracum. It is thought that crystallisation occurs on protein particles in this fluid. Hare (1963) suggests, and it appears reasonable, that the crystal form may depend on the amino acid compositions of these micellae (Plate 5).

Crystallisation from ions in solution is dependent on the partial pressure of CO_2 in the extra-pallial fluid, derived from respiration, which has to be less than 15 per cent for deposition to occur. The enzyme carbonic anhydrase which catalyses the formation of carbonic acid occurs only in a few species and so is clearly not essential to shell formation (Fretter and Graham 1962). Land snails obtain calcium from their food, and aquatic species appear to concentrate calcium and related metal ions from the surrounding water by absorption through the mantle. Fresh-water molluscs never inhabit acid waters deficient in calcium.

Valuable information on shell formation has been obtained from repair in gastropods. Fretter (1952) showed in the snail *Helix pomatia*, by a labelled tracer technique, that calcium is transported from the digestive gland to the area of shell repair. Formation of new shell is rapid and minute crystals of aragonite are laid down in the damaged site within one hour of trauma. Calcite appears after three hours, and in twenty-four hours the surface becomes opaque and hard (Saleuddin and Wilbur 1969). Nevertheless, the repaired area does not present the normal ordered arrangement of crystalline layers. Although the periostracum is generally considered an epidermal product, Abolinŝ-Krogis (1968) believes that in the regenerating shell of *Helix* the matrix protein and calcium salts are derived from the digestive gland. Fretter and Graham (1962) have suggested that the epidermis beneath the calcified shell in gastropods may be incapable of synthesising new matrix protein because its cells are flattened and not like protein-secreting cells. Nevertheless, epithelial cells are capable of rapid changes in synthetic activity in response to changed requirements as in pearl formation. Indeed, it must be remembered that earlier in life these same cells presumably formed periostracum; for if one considers growth in time, cells divided from epidermal cells which form periostracum in the mantle groove must later be displaced under the shell and change to forming prismatic layer and afterwards be displaced further and form nacreous layer. This suggests that changes in secretory activity must occur during the life of the cell.

VARIOUS CHITINISED STRUCTURES

Beta chitin and sclerotins occur in the byssus threads of certain bivalves and also in the gastropod radula, but chitin never occurs in the shells of either gastropods or lamellibranchs in contrast to brachiopod shells, although present in the lower molluscs such as *Chiton*. Chitin occurs in very high concentration in the shells of squids and other cephalopods. In these molluscs, the formative epidermis has sunk into the body, so that in both squids and cuttlefish, although not in *Nautilus*, it is an internal structure.

PECULIARITIES OF THE INTEGUMENT IN CEPHALOPODA

The leathery skin in octopus, squids and cuttlefish is strengthened by cartilage in the dermal connective tissue which gives skeletal support to these often large invertebrates. The cephalopod epidermis is ciliated over the tentacles. The particularly hard beak is probably a mineralised or sclerotised cuticular structure. The interesting pigment cells are discussed in the next chapter.

HEMICHORDATA (ENTEROPNEUSTA AND PTEROBRANCHIATA)

This is now usually given separate phylum status and includes *Balanoglossus* and *Rhabdopleura*. The epidermis is a ciliated epithelium similar to that of nemertines with tapering basal processes dipping into the dermis, and between them are mucous glandular cells and sensory cells. Numerous compound mucous glands occur in the dermis and discharge to the surface (Hyman 1959).

TUNICATA

The Tunicata (Urochordata) include the sea squids (Ascidiacea) and also various planktonic species, some of which form a thick translucent outer test or tunic (Fig. 20). This contains upwards of 5 per cent cellulose chemically similar to that of flowering plants, an unusual polysaccharide to find in animals, although small amounts have been detected in mammalian dermis. The epidermis in ascidians is a single layer of columnar cells, over which lies the test which has a hydrated gel consistency, contains acid mucopolysaccharides, and is crisscrossed with cellulose and protein fibres. Its high content of bound iodine is another peculiar feature (Barrington and Thorpe 1968). The main thickness of the test is not quinone-bonded and is readily digested by trypsin, but there is a thin superficial tougher layer of dense trypsin-resistant sclerotin interspersed with cellulose fibres and without acid mucopolysaccharides. Barrington and Barron (1960) showed that iodine in this region is in the form of diiodotyrosine, possibly involved in quinone linkages. Neither chitin nor appreciable lipids have been found in the ascidian test. Within the test are migrant, probably dermal, amoeboid cells which contain cytoplasmic granules rich in protein-bound sulphydryl groups and also oxidation products of tyrosine required for sclerotisation. Beneath or above the superficial tanned layer there is a discontinuous cellular layer. In some species, blood capillaries loop up between the epidermal cells into the test which gives it the appearance of connective tissue.

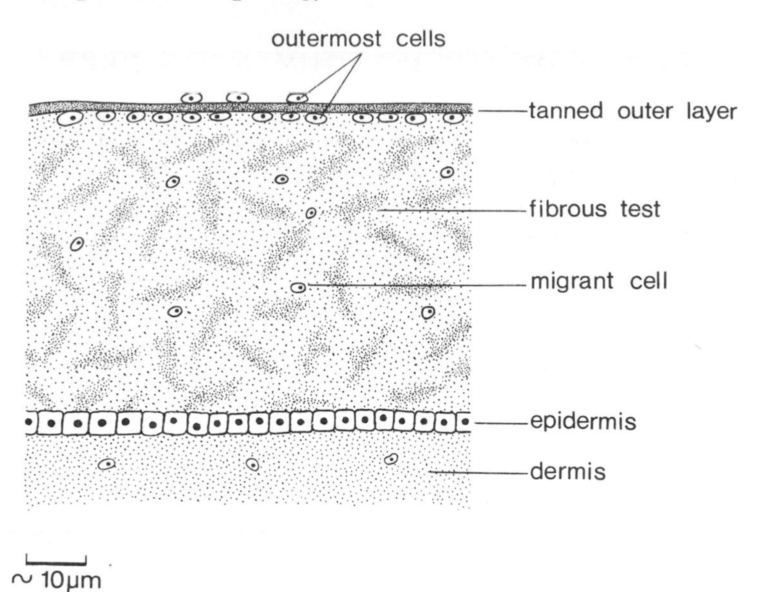

Fig. 20. The tunicate test.

Of the various theories of test formation which have been suggested, only two need be mentioned here: that it is an epidermal secretion, and that it is formed from connective tissue substances that diffuse out between the epidermal cells or are secreted by migrant mesodermal cells. Dilly (1969), from an electron microscopic study in the tadpole larva of *Ciona intestinalis*, believes that the test is probably largely an epidermal exosecretion although the dense surface layer probably receives some material from mesodermal cells, including tyrosine oxidation products utilised for quinone bonding of proteins. Blood capillaries when they occur in the test also react for tyrosine derivatives, and presumably facilitate transport to the outer region when the test is thick.

Thus, the tunicate test appears to be an epidermal cuticle with added material from the dermis. The presence of migrant mesodermal cells above the epidermis is not all that unusual and occurs in snakes and lizards during sloughing as well as in pathological conditions in human skin. Blood capillaries also loop up between the epidermal cells in leeches and some Amphibia.

6

INVERTEBRATE SKIN COLOURATION

The colouration of animals depends on both chemical and physical factors. Pigments in the integument and in deeper tissues such as the blood absorb light of certain wavelengths, while others are reflected and produce skin colour. Reflection and diffraction from opaque surfaces also produce colour (Munro Fox and Vevers 1960).

PIGMENTS

There are many more chemical forms of pigments than occur in vertebrates.

MELANIN

The most common skin pigment is melanin which gives the integument a yellow to black colour. In vertebrates, melanin is synthesised in melanocytes derived from the neural crest, but in invertebrates the origin of these cells is uncertain. The mechanism of melanogenesis is discussed more fully in chapter 15. Melanin is a stable indole quinone polymer linked to protein and is an oxidation derivative of tyrosine. It is often laid down in cytoplasmic bodies, melanosomes, but also occurs outside cells in invertebrates. Slightly different forms produce yellow to black skin tints. Melanocytes later in development may become melanin-storing cells (melanophores). These are found in turbellarians, certain echinoderms, and ctenophores. Frequently melanosomes are phagocytosed by other cells which in this way become indirectly pigmented. Therefore, unless the oxidation stages of melanin formation are observed in cells with this pigment, they cannot be positively identified as formative cells.

SCLEROTINS

In addition to intracellular and extracellular melanin, many invertebrates produce brown pigmented quinone polymers which form cross-linkages in sclerotins. This is a chemical side-track from melanogenesis.

OMMOCHROMES

These are brown, yellow or red granular pigments found only in invertebrates and derived from tryptophan via kynurenine by oxidative condensation. Dopa-quinone formation from Dopa in melanin synthesis is associated with ommochrome synthesis because it acts as an electron acceptor inducing condensation. Unlike melanin, which is highly insoluble, ommochromes are soluble in mineral acids.

COLOUR CHANGES IN CEPHALOPODS

Brown and red ommochromes in separate sets of large chromatophores occur in the cephalopod dermis with, beneath them, iridophores containing crystal plates which produce blue colours by reflected light. The peculiar chromatophores can be rapidly altered in shape by radially attached muscles operated by both excitatory and inhibitory nerves derived from a centre in the brain. Rapid blanching or flashes of colour are produced in a wide variety of tints by changes in chromatophores in different skin sites. When the cells are pulled out and flattened in respect to the skin surface, the skin colour appears dark, but when they round up by elasticity, pigment is clumped and the area between the chromatophores is increased so that the skin is blanched. Camouflage changes to match the background are mediated through the eyes and brain and used to be considered to be reflex actions. In the octopus the mechanism is more complex, for patterns are still produced when the eyes are shielded. Indeed a whole repertoire of possible patterns is remembered in the brain and can be induced by electric stimulation of particular brain areas. The normal computation involves selection. Transient behavioural patterns initiated in the brain also occur, such as to frighten an enemy. The basic pattern is mottled in the octopus (Packard and Sanders 1969).

Chromatophores operated by nerves also occur in leeches, but most invertebrates do not show rapid colour changes.

CAROTENOIDS

These are chemically related to vitamin A (retinol) and occur as a yellow pigment in fats in both invertebrates and vertebrates. They frequently occur in cuticles and produce the orange and red colours of some beetles. In the lobster *Homarus*, a carotenoid is chemically linked to a protein, which in combination forms a dark blue pigment. The change in colour to red in a cooked lobster is due to heat denaturation of the protein with release of the carotenoid, which on its own is red.

PORPHYRINS AND RELATED PIGMENTS

In porphyrins, pyrroles form a ring structure. When combined with iron a haem molecule is produced, and this combined with a protein gives haemoglobin. The latter occurs in the earthworm and in some nematode cuticles. Many annelids and molluscs have porphyrins and chemically related bilins which produce the green colours of some polychaetes. Bilins also occur in sea anemones, corals and molluscs.

PTERINES AND FLAVINES

Pterines are yellow pigments in the cuticles of butterflies and wasps, and are related to folic acid, purines and flavines. A flavine combined with iron gives the red colour of the precious coral *Corallium rubrum*.

BIOLUMINESCENCE

Luminescence of the integument occurs in many invertebrates, and sometimes the luminescent cells are controlled by nerves. It has been most critically studied in insects, in which luminescence is produced by pyrophosphate released from a non-luminous precursor substance, luciferin, synthesised in specialised cells. The chemical reaction utilises energy from the breakdown of ATP and is catalysed by an enzyme, luciferase, in the presence of magnesium ions (Rockstein 1964).

PHYSICAL COLOURS

Iridescence is produced by light interference on reflection from dissimilar surfaces, such as the upper and under sides of thin lamellae of mother-of-pearl. Whiteness is produced by reflection from a regular surface in the absence of pigment. Diffraction of white incident light by a finely ridged or grooved cuticular surface also produces physical colours as in some beetle cuticles.

SUMMARY OF THE INVERTEBRATE INTEGUMENT

Invertebrate skin is involved much more widely in the functions of the animal than that of vertebrates. In addition to protection and sensation which are the basic functions, in many invertebrates it also serves for ciliary locomotion and for transport of food and waste. Apart from arthropods, where the cuticle is thickened as an exoskeleton, the skin is used for respiratory exchanges and, more rarely as in parasitic platyhelminthes and acanthocephala, it is used for food absorption.

Individual epidermal glandular cells, as well as compound glands derived from the epidermis, secrete a variety of fluid substances over the skin surface, the most important of which is protective, viscous or watery (serous) mucus. In other instances, the exosecretion hardens to form a cuticle, and the secretory surface is increased by microvilli. These also occur over food-absorbent surfaces of the intestinal epithelium and in the tegument of tapeworms where increase in surface area is similarly required.

Except in some primitive acoelomates, the epidermis has a defined dermo-epidermal junction provided by a thin basal lamina.

Invertebrate epidermis is, except in rare instances, a simple epithelium one or two cells in depth. The cuticle in different phyla is constructed of different materials. The two most important types are the collagenous cuticles of nematodes and annelids, and the fibroin cuticles of arthropods and some other invertebrates. Chitin is often combined with proteins which gives additional strength. Both the fibroin and collagen may be stabilised by quinone linkages. Various mucopolysaccharides in addition to chitin have been found in cuticles.

Keratin is laid down inside vertebrate epidermal cells and its molecular complexity probably precludes its occurrence as a cuticular constituent. Sulphur amino acids occur in some cuticles and are not confined to keratin. The only conceivable instance of an invertebrate keratin seems to be in parasitic platyhelminthes in structures formed within the syncytium.

A wider variety of pigments is responsible for skin colour in invertebrates than in vertebrates.

7

THE FISH INTEGUMENT

The phylum Chordata will be taken to comprise the Cephalochordata and Vertebrata since the Hemichordata and Tunicata are now usually considered as independent phyla.

Fish in a broad sense comprise a variety of non-tetrapod vertebrates with both jawless and more advanced jawed species arranged in several different classes. For convenience the Cephalochordata will also be considered in this chapter. First the general characteristics of vertebrate skin are considered.

GENERAL FEATURES OF VERTEBRATE SKIN

Even the most primitive living fish, the jawless cyclostomes, have a characteristic vertebrate skin with a stratified epidermis several cells in depth. In lower vertebrates the germinal basal layer is multipotent and gives rise by division to cells which probably secrete mucin and are destined to form the cuticle (cuticularcytes), various glandular cells, and probably sensory receptors. In higher tetrapods the general cell potency is restricted to keratin-forming cells, while glandular cells are confined to the epidermal appendages. Receptor cells in higher vertebrates, although probably derived from the epidermis, are mainly grouped together in the dermis.

Various junctions occur between epidermal cells, the most common form being the *macula adhaerens*: desmosomes responsible for the prickle attachments seen in light microscopy (Fig. 33*b*). Opposing hemi-desmosomes can be readily separated by migrant cells. They also occur facing the basal lamina. Fused plasma membranes, *zonula occludens* (Fig. 26), form belts around cells in epithelia which undergo active transport and also generally occur between superficial epidermal cells. In vertebrates they replace septate desmosomes.

Various kinds of ultrastructural filaments occur in epidermal cells. Tonofilaments in both vertebrates and invertebrates are arranged in bundles attached to desmosomes. Finer filaments are associated with microvilli, and various filaments not obviously associated with any organelle are also seen in electron micrographs. Some of these are prob-

ably restraining or skeletal in function, while others near the plasma membrane are contractile actin filaments (Bereiter-Hahn 1971).

Immediately beneath the epidermis there is a thin basal lamina. In vertebrates the skin connective tissue is always clearly demarcated from the underlying organs by a dense collagenous fascia covering the skeletal musculature and occasionally over skeletal bones.

There is a collagenous dermis (corium) which, except in cyclostomes also contains another fibrous protein, elastin. The dermis in paraffin-processed histological sections is often sub-divisible into a superficial compact layer made up of a fibrous collagen gel and a network of fine elastin filaments, and the deep dermis which appears loosely arranged with coarser collagen fibres. Various mesodermal cells are scattered in the superficial dermis, and calcified scales and teeth are also laid down in this zone and in the basal lamina. The deep dermis contains fewer cells. Cartilage is sometimes present.

Beneath the dermis is the hypodermis (subcutaneous tissue) with a network of collagen fibres which attaches the integument either loosely or firmly to the underlying fascia. The hypodermis contains few to many fat storage (adipose) cells depending on the species. Large lymphatic channels which supplement the blood vasculature occur in this region in most vertebrates but are absent in cyclostomes. The integumental connective tissue is generally well supplied with blood vessels which branch to form capillary plexuses. It functions as a blood storage organ to buffer changes in blood pressure.

Large cutaneous nerve bundles branch in the dermis to form plexuses which contain both cerebrospinal and autonomic fibres. Sensory nerves include myelinated and non-myelinated fibres, and axons penetrate between the epidermal cells. Some sensory nerves connect with specialised receptor cells or groups of cells situated in the epidermis or dermis. Motor nerves supply muscles and some skin glands, but other glands are hormonally controlled.

CEPHALOCHORDATA

This small group is represented only by *Branchiostoma* (Amphioxus) which, while it shows many typical chordate characters, has a much simpler organisation than fishes. It was probably a stock such as this which gave rise to the jawless fish (Hadzi 1963), but another view is that Amphioxus is degenerate rather than primitive.

The skin consists of a single layer of cuboidal cells rich in tonofilaments and with a border of microvilli covered by a thin fibro-proteinous cuticle. Cuticularcytes, glandular and sensory cells are found in the epidermis, and there is a thin dermis (Olsson 1961; Welsch 1968).

AGNATHA

These are the most primitive jawless vertebrates. Extinct Agnatha mostly had a dermal bony armour, but living cyclostomes have a soft skin devoid of scales.

CYCLOSTOMATA

The adult epidermis

In both lampreys and hagfish this is extremely thick (up to 90 μm) and many cells in depth. Germinal cells in mitotic division are confined to the lowest two cell layers. Basically, two kinds of cells occur in the epidermis; cuticularcytes and various larger goblet glandular cells which discharge their contents either into the intercellular spaces or when they reach the skin surface. The most glandular epidermis occurs in the hagfish, which secretes a profuse mucous slime. In addition to individual glandular cells, multicellular mucous glands derived from the epidermis occur in the dermis, and discharge through ducts to the surface. Mucus secretion in the hagfish is under nervous control and is so profuse that if one is placed in a bucket of sea water, the latter is replaced in a short time by jelly, possibly a protective mechanism against ectoparasites.

The cyclostome epidermis secretes a thin cuticle over a border of microvilli (Brodal and Fange 1963).

The larval epidermis

The ammocoete larva which has a ciliated median ventral groove has a much simpler skin more like that of Amphioxus.

Sensory receptors

Groups of elongated pear-shaped receptor cells with sensory filaments occur in the epidermis and are supplied with nerves. Skin photoreceptors have also been described. Touch and probably chemical sense (taste) receptors occur in the circumoral tentacles of the hagfish.

The connective tissue

Absence of elastin fibres is probably a primitive feature. Fibroblasts and prominent melanophores occur in the dermis. The thick hypodermis is packed with large fat storage cells. The dermis and hypodermis contain extensive blood capillary plexuses.

The oral cavity

In all vertebrates this is lined by ectoderm, an extension of the epidermis. It is of particular interest in cyclostomes because the oral cavity is covered

by small horny teeth which are not true teeth, but keratinised epithelial structures. The keratin contains both protein-bound cystine bonds and bound cysteine: phylogenetically it is the earliest definite instance of this protein.

FOSSIL AGNATHA

These primitive fish, together with the jawed placoderms, were the dominant species in Silurian and Devonian seas. The majority of fossil Agnatha had a heavy armour of bony dermal plates. They are divided into several orders. The various ostracoderms, as they are termed, had a carapace of closely fitting, heavily ossified dermal bones over the forepart of the body with free caudal scales. Yet, despite dermal ossification, the skeleton as in modern cyclostomes was cartilaginous. In early fish these calcified dermal plates may have served as protection against water scorpions (eurypterids). The use of bone for skeletal support came much later in evolution and membranous dermal bones long preceded ossification of vertebrae.

Skin ossification probably occurred independently more than once in early vertebrates. The membrane bones in the skulls of tetrapods are similar and were derived from these bones.

The plates in ostracoderms were constructed of several calcified layers. There was a thin basal layer of lamellar bone, then a thick layer of cancellous bone with concentric lamellae around the blood vessels. Above this there was a thin lamellar layer and on top a layer of dentine. Tarlo (1964) traces changes in the cancellous layer in members of the order Heterostraci from the Ordovician to Devonian periods. In the earliest species, the concentric lamellae show no trace of trapped formative cells which presumably withdrew completely including their processes as lamellae were deposited, but in more advanced forms holes occur in the lamellae, originally occupied by the osteocytes which remained in the calcified tissue. Some resorption and remodelling clearly occurred during scale growth as in bones of higher animals.

One Silurian jawless fish *Jamoytius* had a naked integument and showed a number of apparently primitive features. Modern cyclostomes were probably derived from such a group, with Amphioxus possibly a further stage back.

APHETOHYOIDEA (PLACODERMS)

The most primitive jawed fish were the extinct placoderms of the Devonian. Like ostracoderms, they had a heavy armour of ossified dermal plates which in many species formed a cranial carapace while the skeleton was unossified.

A number of different lines of development occurred in the placoderms. The close resemblance to ostracoderm armour is seen in structure of placoderm scales, but dentinous tubercles are generally found rather than a continuous layer of dentine and sometimes these tubercles are free from the underlying scale. One subclass, the Arthrodira, had a hard compact calcified layer over the tops of the scales without any trace of dentine.

CHONDRICHTHYES (ELASMOBRANCHII)

These cartilaginous fish are represented by the living sharks, skates, rays and chimaeras. Dermal ossification had been secondarily lost so that only the placoid denticles remain, related in phylogeny to the tubercles of placoderms. Even these are relatively few in rays and nearly all are lost in chimaeras. The denticles provide the rough surface of shark leather (shagreen). The epidermis is several cells in depth and in addition to cells which are probably cuticularcytes, contains scattered glandular goblet cells and sensory cells. The dermis is particularly dense and fibrous and has a two-layered structure. A cuticle has not been described but probably occurs.

DERMAL DENTICLES

The placoid denticle closely resembles a mammalian tooth in both structure and development (Fig. 21). This is the only type of fish scale which continues to be formed *de novo* throughout life, so that the number in the skin increases year by year. Completed denticles do not grow in size and they eventually get worn down or torn out through wear and tear. This is in contrast to the continuous growth in size of other types of fish scales. Fish do not stop growing as do terrestrial vertebrates when they reach maturity and so the skin surface continues to enlarge year by year. In consequence, increase in number or size of scales is necessary to maintain density in the skin.

In denticle development a papilla composed of a group of mesodermal cells with dendritic processes appears in the dermis. These cells probably secrete the collagen matrix on which crystallisation occurs, mainly of calcium phosphate as hydroxyapatite with some calcite. A cone of crystalline calcium salts is thus laid down with the formative cells against the inner calcified surface with only their processes penetrating the calcified material: the characteristic feature of dentine. Both dentine and bone contain the same salts but in bone the cells become surrounded by calcified tissue and not just their processes.

The soft pulp in each denticle is richly supplied with blood capillaries and is the most vascular region of the elasmobranch dermis. Sensory

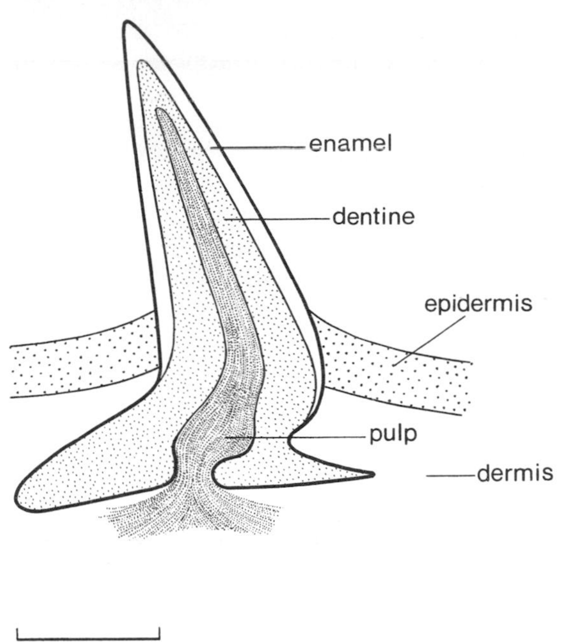

Fig. 21. Elasmobranch placoid denticle.

nerves also end in the pulp cavity. As dentine formation proceeds, the cone reaches up towards the epidermis. The epidermal basal layer (enamel organ) then grows down over the dentine cone and deposits over the denticle a thin non-cellular layer of very hard crystalline enamel with a much lower organic content than either bone or dentine.

As the denticle grows upward, the point eventually pierces the epidermis, but it remains firmly held in the dermis by its expanded base. Differences occur between the enamel and dentine of fish denticles and of mammalian teeth, but seem of relatively minor significance.

ACTINOPTERYGII (GANOID FISHES)

The evolution of these fishes with a bony skeleton shows a progressive reduction in the dermal scales from the ossified scutes of extinct palaeoniscoids to modern teleosts, which have only weakly calcified scales. The palaeoniscoid scale was rhomboidal in shape and a thick bony structure (Fig. 22). It had a complex layered arrangement with, from the underside upwards, a thick layer of compact lamellar bone (referred to in old accounts as isopedine), then a thin layer of cancellous bone followed by a layer of dentine, and on the upper surface a thick layer of ganoin, a calci-

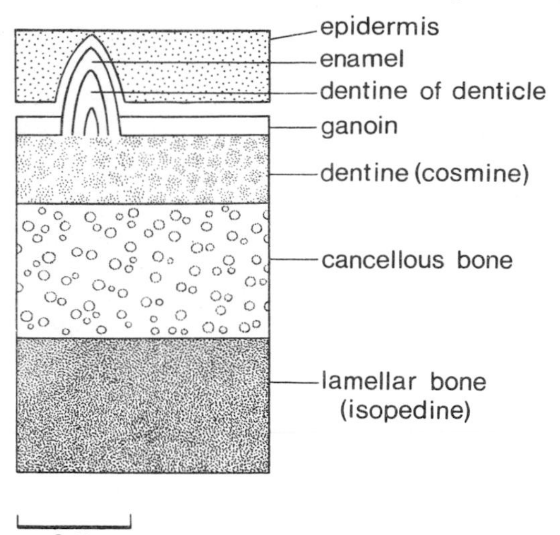

Fig. 22. Diagrammatic vertical section through generalised ganoid scale.

fied non-cellular material without canals and with a low organic content. To this extent ganoin resembles enamel, although the two substances differ in their modes of formation. Enamel is exposed at the skin surface when a tooth or placoid scale erupts, but ganoin always remains subepidermal. A consequence of this is that once enamel is formed, it cannot be increased in depth or repaired. Ganoin, in contrast, continues to be deposited as new lamellae throughout life. Growth in size of a ganoid scale occurs also by deposition of lamellar bone on the lower surface as well as laterally and there is some resorption and remodelling. Canals for blood vessels pierce both the upper and lower scale surfaces, which shows conclusively that fossil specimens were subepidermal.

It has been suggested that the epidermal basal cells may secrete a protein into the basal lamina which acts as a nucleus for crystallisation of dermal scales, so that both dermis and epidermis may be involved.

The dentine in ganoid scales used to be termed cosmine because canals are branched instead of straight.

LIVING GANOID FISHES

Simplified ganoid scales occur in *Polypterus* of African rivers. Kerr (1952) showed that the skin contains rows of slightly overlapping dermal scales bound together by collagen fibres and the scales are separated from the epidermal basal layer by only a thin membranous connective tissue. This supports the idea of epidermal participation.

Polypterus epidermis contains non-glandular cells interspersed with

glandular goblet cells. The scales have a thick ganoid layer over a layer of dentine, beneath which is lamellar bone, but the cancellous bone of palaeoniscoids has been lost.

The gar pike *Lepidosteus* of the Great Lakes of North America has a similar type of skin but with prominent deeper-seated scales (Kerr 1952). On the upper surface is a layer of ganoin and the dentine layer is broken up into separate denticles. Some of these remain beneath the ganoid layer while others occur isolated from the scale. Some denticles pierce the epidermis as in elasmobranchs. It will be remembered that similar break-up of the dentine layer into tubercles occurred in placoderms. In the gar pike, new denticles are formed throughout life, as in sharks.

GROWTH OF THE GANOID SCALE

In development, a group of osteoblasts first lays down a bony plate in the superficial dermis and growth then continues on all sides. Ganoin is later laid down and continues to be deposited in thin layers throughout life. Its lamellar arrangement suggested to Kerr (1952) that cyclical deposition of ganoin may be hormonally controlled.

The nearest living relative of the gar pike is the bow fin *Amia*, also a fresh-water species from North America. This has undergone a more extreme reduction of the scales which are merely collagenous plates containing bony particles. Both ganoin and dentine have been lost.

In *Amia* the epidermis forms a thin cuticle (Whitear 1970), which is probably true of other ganoid fishes.

The sturgeon *Acipenser* and paddle fish *Polyodon* are other related fishes. Sturgeons have large bony scutes which are devoid of ganoin and dentine, but the paddle fish has vestigial denticles similar to those in the gar pike.

THE DERMIS IN TELEOSTS

Teleosts include the majority of living fishes. The dermal scales are reduced to thin flexible collagenous plates in which calcium salts are deposited. Ganoin never occurs, while dentine and bone occur in very few species.

Cycloid and ctenoid scales

Teleost scales from their surface appearance are divisible into two types. The more primitive species have cycloid scales with a smooth posterior border, as in the cod, and these fish also have soft fin rays. More advanced teleosts, such as the perch, have ctenoid scales in which stiff fibrous spines extend from the posterior border of each scale and the fins have similar dermal spines, sometimes calcified as in the stickleback.

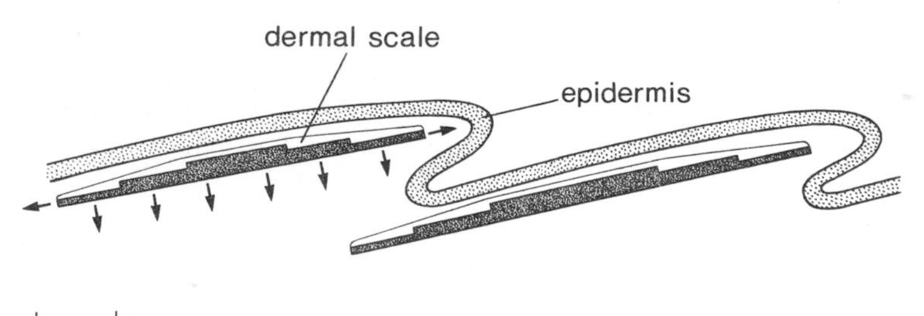

Fig. 23. Diagrammatic vertical section through teleost skin with two-layered scales. Arrows show growth.

A few teleosts have other types of scales. Thus, the snipe fish has denticles, while the sea horse *Hippocampus* has a rigid dermal skeleton of bony plates. The porcupine fish has bony spines and the giant sunfish has a layer of dermal cartilage.

Composition of the teleost scale

The fibrous component of the scale is collagen on which calcium phosphate as hydroxyapatite together with carbonate are deposited. The scales contain 41–84 per cent protein (Oosten 1957), and are surrounded by an elastin framework.

Scale development has been studied in the trout and commences as a papilla of cells which arrange themselves into two layers (the upper layer possibly of epidermal origin) with the scale laid down between them. In many species there is a distinct two-layered scale structure with the upper part more strongly calcified (Fig. 23). Others, such as the goldfish, have the whole depth of the scale calcified (Oosten 1957). Scale growth continues throughout life from the margin of the upper surface and over the whole of the lower surface. Seasonal growth in temperate waters results in the characteristic annual growth rings due to the step-like deposition in the lower part of the scale. Scale rings are used by fishery biologists to determine the ages of fish, but are not always reliable.

In most teleosts the scales overlap as they grow larger and they are always covered with a thin layer of dermis and epidermis. Replacement occurs only in damaged areas. In some species they are easily ripped out as in the herring *Clupea*, which because of this is difficult to keep alive after capture for aquaria purposes. In histological sections, the teleost dermis is arranged as a superficial layer containing the scales and a deep layer with larger collagen fibres mostly parallel with the skin surface. Sometimes fat is stored in the deep connective tissues as in the herring.

Blood vessels in teleosts

The skin in fish is an accessory respiratory surface and in consequence requires a good blood supply. The cutaneous vasculature has been examined in a number of species (Jakubowski 1960). Arteries to the skin break up to form capillary plexuses which are drained by veins and lymphatics. The superficial dermis around the scales is most strongly vascularised. Each scale has its own arteriole and venule with an overlying capillary plexus, but capillaries never enter the epidermis.

Teleost epidermis

This contains cuticularcytes as well as glandular and sensory cells (Kann 1926). In electron micrographs the cuticularcytes contain masses of filaments. In addition to large mucous cells, peculiar club cells occur, the secretion of which seems to alert other fishes to danger. The basal cells are not distinct in appearance from the cells above as they are in higher tetrapods.

The cuticle

In most bony fishes the thin fragile cuticle is often lost during histological processing, but it is thick in *Hippocampus*. Whitear (1970) found a cuticle in all the specimens of fish skin which she examined in a variety of genera, and probably it is of general occurrence, but the chemical composition has not yet been determined (Plate 6). The cells underlying the cuticle have prominent microvilli and a granular endoplasmic reticulum generally associated with protein exosecretion. The cuticular structure varies from species to species. Sometimes ultrastructural cytoplasmic vesicles contribute to the cuticle, and often fibrous material appears to pass through the membrane lining the microvilli. The plasma membrane itself may be incorporated into the cuticle.

In some species the cuticle is clearly ecdysed at intervals, assisted by the violent discharge of mucus from the epidermal goblet cells (Gilchrist 1920). Ecdysis presumably occurs regularly in fish despite the fact that it has only rarely been described. Indeed, it must be essential to allow for the continued expansion of the skin surface during growth.

Unfortunately little is known about cyclical changes in fish epidermis. Nevertheless, ecdysis in nematodes and arthropods and the cyclical changes in higher vertebrate epidermis suggest that ecdysis in fish is probably also hormonally controlled, as is true of breeding tubercles.

A thin cuticle is continued over the oral cavity and gill surfaces which are ectodermal in nature.

Electric organs

In the electric catfish, certain epidermal cells are modified as electric storage plates, but other fish which generate electric currents utilise modified muscle cells.

Keratinisation

This process, also called cornification, is uncommon in fish and occurs only in restricted sites. Probably the horny lips around the sharp jaws of herbivorous fresh-water species are keratinised, and there is evidence of keratinisation in the breeding (nuptial) tubercles of some fresh-water and inshore marine teleosts. The rough surface provided enables individuals to maintain close contact during spawning. Wiley and Collette (1970) describe three different kinds of breeding tubercles. The first is composed of not obviously keratinised epidermal cells. In the second type a hard conical keratinised epidermal cap is formed which resembles a plant thorn (Plate 7). The cells become flattened as they are keratinised and die, and the nuclei are either lost or become shrunken (parakeratotic). Evidence for keratinisation comes from histological appearance and staining reactions, and bound cysteine has been detected. The present author found bound phospholipids in *Leuciscus* keratinised tubercles, often seen in cornified cells. The third type of tubercle is the contact organ which, although similar in function, is quite different and is formed of calcified dermal outgrowths from scales or fin rays.

Epidermal tubercles are usually sex-limited to males or are better developed in males. They form during the breeding season and are afterwards moulted. In histological appearance some of these tubercles appear as discrete keratinised appendages (Wiley and Collette 1970). They occur over the head and body.

The teleost gill epithelium

The thin gill epithelium provides a large surface area for respiratory exchanges and the connective tissue is highly vascularised. In addition to typical cuticularcytes and mucous cells, eosinophilic (chloride) cells occur, not found in the skin. These may be concerned with salt regulation, although recently doubt has been shed on their function (Doyle and Gorecki 1961). Certainly the gills actively absorb salt in fresh-water fish and secrete it into the sea water in marine species, but there is no direct experimental evidence that these cells are involved.

Skin innervation and sensitivity in teleosts

The skin in all fishes is well supplied with nerves. Whitear (1971*a*) found in the minnow both deep dermal and sub-epidermal plexuses, as in higher

vertebrates. Both free nerve endings and nerves to spindle-shaped sensory receptor cells occur in the epidermis.

Fishes are sensitive to touch and probably also to pain stimuli as well as to temperature. Certain epidermal receptors (Plate 8) resemble mammalian taste cells and are chemoreceptors (Whitear 1971*b*).

The lateralis system of fishes

The lateralis sensory system (Dijkgraaf 1963) is developed only in fish and in the more aquatic Amphibia. It is supplied by cranial nerves VII, IX and X. There are two different types of organs. The first is the ampullary organ of elasmobranchs, ganoid fishes and lung-fishes. This consists of pits in the epidermis lined with receptor cells with a jelly of high electric conductivity in the cavity. Ampullary pits are scattered over the head region and are electroreceptors sensitive to electric fields so that fish will orientate in an electric current.

The second type is the neuromast system which registers vibration and water movement. Isolated neuromasts occur over the head in cyclostomes, jawed fishes and the more aquatic amphibians. Neuromasts contained in lateral-line canals along the sides of the body occur in most fishes. The neuromast consists of a cluster of pear-shaped receptor cells surrounded by supporting cells. Sensory filaments from the receptors project into the cupula jelly over the neuromast. These organs develop along lines on the head and trunk. In development, the epidermis sinks in and closes over to form a canal with occasional openings to the surface. The labyrinth of the inner ear in mammals is derived from this type of system. Indeed, there is a close connection between lateralis organs and the inner ear of higher vertebrates, so that it is termed the acoustico-lateralis system.

CHOANICHTHYES

Extinct members of this class had large dermal cosmoid scales composed of a basal layer of lamellar bone, above it a layer of cancellous bone, then a layer of cosmine-type dentine, and on the upper surface a layer of harder material called vitrodentine (Fig. 24). Growth of the scale was by addition of new lamellar bone underneath, but not over the upper surface.

The class is represented by three species of lung-fishes (Dipnoi) and a single coelacanth species of the subclass (Crossopterygii) found off South-East Africa and Madagascar. This fish, *Latimeria*, shows a reduction in the dentine layer compared with the thick cosmoid scales of extinct species, and it has bony scales studded with denticles. The epidermis in both lung-fishes and in *Latimeria* contains cells which are probably cuticularcytes and numerous mucous goblet cells (Pfeiffer 1968). Modern lung-fish have completely lost the dentine layer and the scales are reduced to thin plates

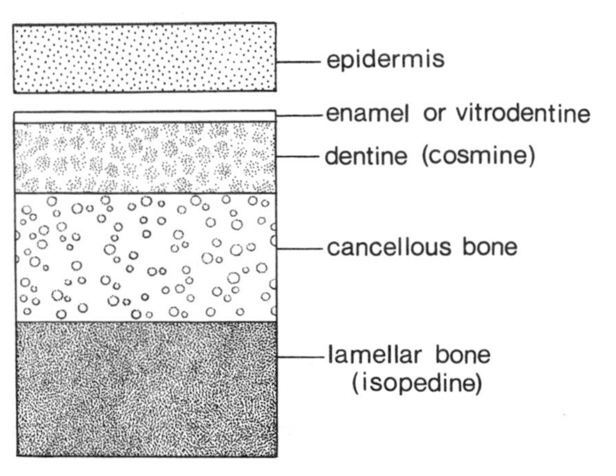

Fig. 24. Diagrammatic vertical section through generalised cosmoid scale.

made up of an outer calcified and inner fibrous layer (Kerr 1955; Kitzan and Sweeny 1968).

Thus, a similar reduction in scale calcification occurred independently in teleosts and lung-fishes. The ganoin and dentine layers were the first to go, and eventually calcified spicules were all that was left of the original bony structure. Elasmobranchs in contrast lost their ability to form bone but retained dentine although even the fibrous component of the ancestral scales which bore these denticles is lost.

SKIN COLOUR IN FISHES

Pigment-containing chromatophores of fishes are located at various levels in the dermis, and the epidermis is always unpigmented. The latter contains migrant unpigmented Langerhans cells and histiocytes. The skin colour varies widely from black or brown to contrasting bright colours and silver in different species. The dorsum is usually darker than the underside, useful for camouflage.

The brown to black pigment, melanin, is contained in melanosomes in large melanophores situated at a lower level in the dermis than other chromatophores. As the skin surface area increases, new chromatophores are added by cell division.

Xanthophores, which produce yellow to red colours, contain pterines and carotenoids. In the goldfish, pterines which produce the orange colour are contained in cytoplasmic bodies analogous to melanosomes. Carotenoids occur in small cytoplasmic vesicles or in fat deposits. The third type of cell is the iridophore, which contains thin plates of crystalline guanine, a nitrogenous material. Reflection of light from the iridophores gives the skin a silvery colour.

While many species have fairly constant skin colour patterns, others show active colour changes produced by a change in distribution of pigment within the cells, as in cephalopods. However, in vertebrates these changes are produced by contractile fibres within the chromatophores and not by external muscles. Dispersion of pigment in the outer cytoplasm and dendritic processes darkens the skin, while clumping of pigment around the nucleus makes it lighter in colour. The silver appearance is most marked when guanine is dispersed. Combinations of pigment changes give intermediate colours (Fujii and Novales 1969).

The most complex colour changes occur in the plaice and other members of the Pleuronectiformes (see chapter 14). However, less is known about the neurological processes than in the octopus, but they appear as complicated with a set of possible patterns stored in the brain. These flatfish are bottom dwellers, and change colour within a few seconds to match the colour and even pattern of the substratum, which they view before they come to rest. In comparison, hormonal changes may take twenty hours or more (Bagnara and Hadley 1969; Greenwood 1963). Deep-sea fish are often luminescent and all are deeply melanotic as distinct from most cave-dwellers which are unpigmented. This paradox has not been explained.

SUMMARY

Fish have a stratified epidermis made up of cuticle-secreting cells, conveniently termed cuticularcytes, glandular goblet cells of various kinds, and sensory receptor cells. Occasionally, compound glands are present. These various skin glands produce mainly mucous and lipoidal secretions. A thin fibrous cuticle is probably always present in fish, and although ecdysis has been described in only a few species, it must occur at intervals to allow for growth.

Keratinisation is unusual in fish, but occurs in the jaws of cyclostomes and breeding tubercles of teleosts.

The superficial dermis contains prominent scales which are fibrous and often calcified. In the early ostracoderms and placoderms, the scales contained a primitive form of bone together with a covering layer of dentine. Only denticles covered by a thin layer of enamel remain in elasmobranchs.

In modern bony fishes, the scales are mostly reduced to wafer-like structures which are much less strongly calcified than in ancestral forms. The first to go was bone and then dentine.

8

THE SKIN OF AMPHIBIA

ADAPTIVE CHARACTERISTICS

No amphibian is completely adapted to terrestrial life. The larval stages are always aquatic and many adult tailed species (Urodela) rarely, if ever, venture on land, while some fully aquatic forms continue to breathe by external gills. Even true salamanders inhabit moist places and return to water for breeding. The tail-less Anura are in various degrees dependent on water. A few are aquatic like the clawed toad *Xenopus*, while the frogs in Lake Titikaka never leave the deepest water. In contrast, desert toads survive arid conditions by hiding under stones or in holes during the day and store water in the urinary bladder. However, they are unable to withstand the desiccating effect of prolonged sunlight.

Amphibian epidermal cells are able to actively transport sodium and chloride ions back into the dermis and so prevent leakage of body fluids, a process which was lost in higher vertebrate integuments.

The skins of larvae and of adult neotenous Urodela show clear morphological differences from metamorphosed individuals. In tadpoles and neotenous urodeles, the epidermis is unkeratinised and covered with a thin cuticle as in fish. Profuse watery mucus is secreted by most amphibians, but the majority of glandular cells are contained in unique multicellular alveolar glands. The skin, as in fish, is an important accessory respiratory surface, which restricts keratinisation. A continuous outer layer of flattened, cornified cells nevertheless occurs in all adult Anura and in metamorphosed Urodela.

THE EPIDERMIS

THE LARVAL CONDITION

Newly hatched tadpoles have a ciliated epidermis only one cell in depth which soon becomes stratified. Cilia are lost after some six days in newt tadpoles but persist for up to six weeks in frogs. Urodele tadpoles have in the epidermis, in addition to typical cpithclial and glandular cells, larger, oval-shaped leydig cells, which appear secretory in function (Kelly 1966). In larval newts, leydig cells make up the bulk of the epidermis but vanish at metamorphosis. After cilia are lost, the epidermis secretes a thin cuticle

but never a horny layer, although cornification occurs over the gums and in the rasping horny teeth of older carnivorous tadpoles. These teeth are similar to those in cyclostomes and are not true teeth but keratinised structures. The larval skin is much more melanotic than the metamorphosed integument.

ADULT EPIDERMIS OF NEOTENOUS URODELA

Persistent larval forms such as *Necturus* and *Proteus*, which as adults are fully aquatic, retain an epidermis similar to that of the late larval stage with leydig cells. In *Necturus*, small mucous goblet cells occur, especially around the mouth. The epidermis develops a thin cuticle and the only sites which become keratinised are the tips of the digits.

METAMORPHOSED ADULT EPIDERMIS

More aquatic species

Some metamorphosed Urodela and Anura spend most of their lives in water. This is true of newts (*Molge*) and hellbenders (*Cryptobranchus*), but they nevertheless have a completely keratinised epidermis. In *Cryptobranchus*, a genus which includes the largest living amphibian, the giant salamander of Japan, the stratum corneum is up to six cells in depth, but in most amphibians it is only one cell deep. Nevertheless, despite a stratified horny layer, the skin remains a major respiratory surface, which is made possible because its surface area is greatly increased by folds, and blood capillaries loop up between the epidermal cells to just beneath the horny layer. Epidermal blood capillaries are found in some other aquatic urodeles and in the epithelium covering the external gills of many tadpoles and persistent larval species (Noble 1931). Leydig cells of urodeles are lost at metamorphosis.

More terrestrial species

The adult epidermis in salamanders and in the Anura is essentially similar (Fig. 25).

The outermost epidermal cells and larval cuticle are sloughed at metamorphosis and the cells underneath secrete a new cuticle and become keratinised as they reach the skin surface. Owing to its fragility, the cuticle is often lost during histological processing, but was demonstrated over the horny layer in adult frog skin by Whitear (1970). Therefore cuticularcytes become keratinocytes.

The adult epidermis contains peculiar flask-shaped cells of unknown function which are particularly rich in mitochondria and so are metabolically active. These cells lie in the mid-region of the epidermis, but as they

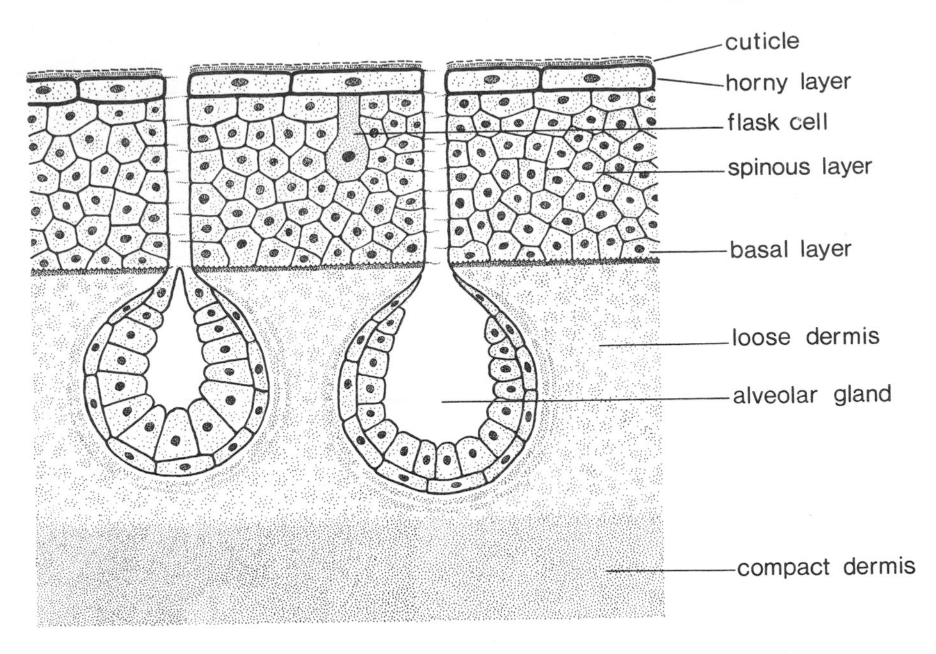

Fig. 25. Diagrammatic vertical section through amphibian skin.

are firmly attached to the horny layer their necks get compressed by surrounding cells. When the horny layer is moulted, flask cells are often lost with it but are soon replaced. Although glandular secretion is mainly confined to the large alveolar glands, probably all epidermal cells are to some extent mucogenic in Amphibia. We see a radical change in epidermal organisation continued in reptiles, birds and mammals, with glandular cells increasingly removed to specialised appendages. The unkeratinised epidermal cells are held together by simple desmosomes, and animals killed with chloroform often show intercellular oedema with taut prickles and dilated spaces between the epidermal cells. The basal cells are not morphologically distinct from cells above. As keratin-forming cells (keratinocytes) are pushed upwards towards the skin surface, they become more flattened. Cornification is sudden, cells beneath appearing quite different from the horny layer (Parakkal and Matoltsy 1964; Spearman 1968*b*).

The horny layer (Plate 9) is a sheet of dead flattened keratin-filled cells which always retain weakly stainable remnants of nuclei. A feature of both

keratinised and unkeratinised species is that membranes between adjacent superficial cells are fused to give tight waterproof junctions, *zonulae occludens* (Farquhar and Palade 1965). The prekeratin protein is elaborated in the prickle cell layer which is rich in ribosomes and RNA required for protein synthesis. The final stages of keratinisation occur in the transitional cells. Stabilisation against protease enzymes is produced by cross-bonding with disulphide linkages and hydrogen bonds. The partial autolysis of cell organelles then occurs by release of lysosomal hydrolases. Neighbouring transitional cells are at first in different stages of keratinisation and the process is not completed until the old layer has been sloughed. This differs from higher vertebrates in which keratinisation in neighbouring cells is synchronised. Keratinisation is closely geared to rate of sloughing. Cystine is concentrated as a thin shell in and under the plasma membrane which appears dense in electron micrographs, and much less occurs in the interiors of the cells which presumably have a different type of keratin bonding.

The cornified cells also contain bound cysteine which is not all oxidised to cystine, and they are rich in keratin-bound phospholipids. Partially-denatured DNA is responsible for the weak nuclear staining reaction with haematoxylin. Usually the nuclei remain large but are sometimes pyknotic. Degraded mitochondria occur together with filaments thought to be alpha keratin (Rudall 1947).

THE SLOUGHING PROCESS

Moult and replacement of the horny layer occurs intermittently in all species. Newly metamorphosed toads (*Bufo*) slough once every three days, and adult toads about once a week. Ambient temperature is the most important external factor in determining frequency. Thus, it occurs much more often in hot weather than at 5 °C (Larsen 1973). Cell loss from the surface is replaced by cell division below so that epidermal thickness does not change. In toads, the horny layer splits along a dehiscence line on the back which probably lacks fused junctions. Sloughing is facilitated by puffing up the body with air and by the animal rubbing the skin against stones or other objects. Body movements also loosen the old horny layer by rupturing the desmosomal junctions with the new horny layer underneath. Lysosomal bodies (membrane-coating granules) in the transitional cells contain hydrolases which are exosecreted into the space between the horny cells and may help break down weak junctions. Sloughing is possibly helped by secretion of mucus between opposing keratinised layers. The animal finally draws itself out of the horny layer which it normally eats. The new skin is at first shiny and glistening.

Despite the presence of fused lateral junctions (Fig. 26) and bound

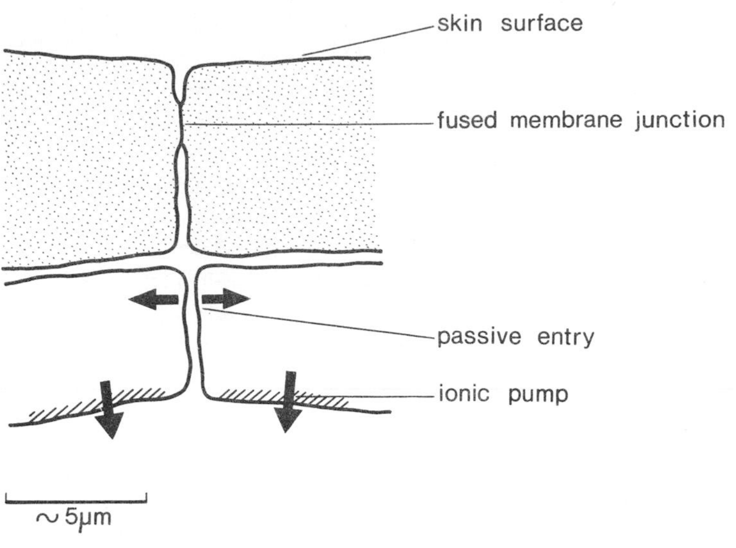

Fig. 26. Adjacent superficial epidermal cells in Amphibia showing fused junction. Arrows show active transport of sodium and chloride. In keratinised species this occurs in the cells beneath the horny layer (after Farquhar and Palade).

hydrophobic lipids in the cells, the amphibian horny layer is much less efficient a barrier to water movement than the thicker layer of higher vertebrates.

Periodic sloughing of superficial epidermal cells occurs in fish and also in unkeratinised urodeles (Dawson 1920). This process was probably originally associated with cuticle ecdysis and when the outermost cells became keratinised, moult of the horny layer was superimposed on the older mechanism. Moult of keratinised breeding tubercles occurs in fish.

Sloughing of the amphibian horny layer is much simpler than in reptiles and does not involve cyclical changes in the keratin, as occur in lizards and snakes. The essential feature is the presence of weak junctions which rupture along the plane of separation. In hypophysectomised toads desmosomal rupture fails to occur and the cells pile up, layer upon layer.

Variation in the horny layer

Although in the frog and salamander the horny layer is only one cell in depth and some 5 μm thick, it is occasionally much thicker, as mentioned in the hellbender. In the sensory tubercles (warts) of toads, superimposed horny cells pile up to 60 μm in depth.

Claws

These occur over the tips of the toes in many Anura and are formed of hard, flattened keratin-filled cells.

THE SKIN GLANDS

Large alveolar glands occupy most of the superficial dermis, and in the frog are packed close together (Fig. 25). The merocrine secretory product is passed from the glandular cells into a large lumen whence it reaches the skin surface by a short duct. Two basic types of gland are recognised: the mucous gland with hyaline secretory cells and the poison gland with granular cells. The mucous glands are the chief source of the watery mucus which gives the skin its moist feel, important on land for it to function as a respiratory surface. However, not all amphibians have moist skins and many toads have dry skins, although they retain the ability to secrete mucus. In these animals, buccal respiration is more important.

The poison glands discharge their contents intermittently when they produce a copious milky solution, as when the skin is irritated by some toxic agent, or if the animal is attacked by a predator. Their function appears to be to deter predators which find the taste unpleasant. Sometimes the secretion is frankly poisonous, as in some toads, but it does not deter grass snakes. A detailed study of these glands would possibly reveal a wide variety of toxic products.

In *Necturus*, the poison glands have an outer sheath of smooth muscle which, together with the secretory cells, are supplied with nerves. Reflex contraction of the muscle causes sudden discharge of the secretion. Mucous glands which show continuous secretory activity are also controlled by nerves but do not have a muscular sheath (Dawson 1920).

MIGRANT CELLS IN THE EPIDERMIS

This term is used for a number of different cells which actively enter the epidermis from below by way of the intercellular spaces.

LANGERHANS CELLS

These unpigmented dendritic cells have a different electron microscopic appearance to melanocytes and the current view is that they are unrelated in development. Langerhans cells occur occasionally scattered in the epidermis. Their function, as in other vertebrates, is unknown.

MIGRANT BLOOD CELLS

Leucocytes and histiocytes frequently wander up into the epidermis, as in other vertebrates.

THE DERMIS AND HYPODERMIS

Beneath the epidermis is a thin basal lamina. The superficial dermis contains collagen and a network of elastic fibres together with acid mucopolysaccharides. This glandular region is separated from the deep dermis in frogs and toads by a thin homogeneous non-cellular mucopolysaccharide- and calcium-rich layer (the figures of Eberth) which stains dark blue with haematoxylin. This layer, which lies just beneath the skin glands, is of undetermined function, but it is not found in other vertebrates. It occurs only in the more terrestrial Anura.

The deep dermis in fixed sections contains coarse collagen fibres, and occasionally bundles of smooth muscle run up vertically to the epidermis.

The superficial dermis contains more cells than the deep dermis or hypodermis. They include fibrocytes and macrophages.

HYPODERMIS

The hypodermis is loosely arranged with large collagen fibres around prominent lymph spaces. There is little subcutaneous fat.

BLOOD VASCULATURE

Related to its respiratory function, amphibian skin is extremely vascular. It has already been mentioned that in some species blood capillaries loop up into the epidermis. Arteries to the skin break up to form capillary plexuses in the superficial dermis, and deeper plexuses also occur. Blood is drained by veins and the numerous lymphatics.

DERMAL OSSIFICATION

Dermal ossification as in fish occurred in the fossil Stegocephalia which had bony scutes, but in living Amphibia small bony nodules in the dermis are confined to the worm-like coecilians (Gymnophiona) and a few species of Anura, such as the horned toad, with projections on the head. True teeth derived from the denticles of fish are confined to the jaws of adult Amphibia.

INNERVATION

Most of our knowledge of cutaneous nerves in Amphibia comes from the early study of *Necturus* (Dawson 1920). In this animal nerves form a plexus in the hypodermis and myelinated fibres pass upwards to supply the skin glands and smooth muscle. Nerve fibres often run for some distance just beneath the epidermis before turning upwards to penetrate between the

epidermal cells where they lose their myelin sheaths and have tapered endings. Sensory receptor cells occur in the warts of toads.

All amphibian larvae and adult perennibranch urodeles are equipped with neuromast organs. These are lost at metamorphosis in most frogs and toads, but are retained in *Xenopus*.

The neuromasts in Amphibia are similar in structure to those described in fish, but are more superficial and are not placed in canals. They occur in the epidermis of the head and flanks (Dijkgraaf 1963).

SKIN COLOURATION

MELANOCYTES AND EPIDERMAL MELANIN

While many melanocytes remain in the dermis and function as melanophores as in fish, others enter between the epidermal basal cells. Epidermal melanocytes are much smaller than the dermal melanophores although both are derived from the neural crest. The epidermal melanocytes contain tyrosinase for melanin synthesis. Some of the melanosomes in the dendritic processes are phagocytosed by epidermal cells, and melanin then forms caps over the nuclei.

Melanin has two functions in the epidermis: camouflage colouration and a screen against ultraviolet light which, particularly in land animals, might damage the living cells. In the brown patches of skin in the common frog, the epidermal cells contain prominent melanin granules which show up in histological sections even without the use of silver techniques.

More melanosomes are formed than are transferred to epidermal cells and the surplus is retained in the melanocyte which continues to function as a melanophore. Melanosomes in the epidermal melanophores, although unaffected by melatonin, expand into the dendritic processes in response to MSH.

DERMAL CHROMATOPHORES

Amphibia have in the superficial dermis the same three types of chromatophores as in fishes. The deepest are the large melanophores. Just above them, and at times covered by the expanded processes of the melanophores, are dendritic iridophores. The most superficial chromatophores are the xanthophores which contain carotenoids and function as a yellow filter to reflected light. Two types of skin colour change occur: rapid physiological change and a slow morphological change due to melanogenesis.

PHYSIOLOGICAL COLOUR CHANGE

Ability of rapid colour change is first developed just before metamorphosis (see chapter 14; Bagnara and Hadley 1969; Montagna and Hu 1967). Melanin is dispersed in melanophores by MSH, but iridophores respond more slowly in the opposite manner by contraction of the guanine platelets around the nucleus. At the same time they are shielded from incident light by processes of the melanophores. The effect of this is to darken the skin, but the basic pattern is unchanged. Colour patterns in amphibian skin are produced by differences in the distribution of melanophores and to a lesser extent of xanthophores. Iridophores are much less important. Xanthophores in the absence of melanophores produce the orange spots in the spotted fire salamander.

On a light background MSH is not released, but cholinergic nerves to the melanophores are stimulated which induces clumping of melanosomes around the nucleus. Because the dendrites and peripheral cytoplasm become free of melanin, the melanophores appear contracted. Again, the iridophores respond in the opposite way with expansion of guanine into the dendrites. The iridophores are now exposed to incident light and form a reflecting surface. The combined effects of melanophores and iridophores is to lighten the skin colour. Hypophysectomy removes the stimulation by MSH so that the melanophores contract and the iridophores expand, which gives hypophysectomised tadpoles a silvery appearance. The innervation of iridophores is probably by the action of acetylcholine, but the nerve endings have not yet been found as they have for melanophores. Colour changes in Amphibia are not as rapid as in flat-fish or in some lizards, and they show little ability to simulate background patterns.

MORPHOLOGICAL COLOUR CHANGE

In addition to affecting the distributions of melanosomes and guanine platelets in the cytoplasm of chromatophores, MSH stimulates melanogenesis in response to sunlight, again acting through the retina and pituitary gland.

SUMMARY

The skin in Amphibia shows features in common with both fish and higher tetrapods. Tadpoles and adult perennibranch tailed species have an unkeratinised epidermis as in fish, but more terrestrial metamorphosed Amphibia have a complete stratum corneum, which probably helps to waterproof the epidermis through the keratin-bound phospholipids present. The horny layer is only one or two cells in depth in most species and the moist skin is profusely supplied with mucous glands, important in

its function as a respiratory surface. The dermis is well supplied with blood vessels, and blood capillaries sometimes penetrate the epidermis between the cells, which facilitates the exchange of oxygen and carbon dioxide. Some species develop locally a relatively thick horny layer as in the horny warts of toads which are sensory structures. Sloughing of the horny layer occurs at frequent intervals. This process appears to have been superimposed on the more primitive sloughing mechanism of the superficial epidermal cells and cuticle in fish. A cuticle also occurs in keratinised Amphibia but ceases to have any functional importance.

Rapid colour changes can be effected as in fish, but simulation of patterns is less exact.

9

THE SKIN OF REPTILES

A WATERPROOF COVERING

Reptiles were the first vertebrates to become fully adapted to life on land. This they achieved in three ways: they are not dependent on water for their early development, the kidneys are more efficient, and the skin has ceased to be a respiratory organ. In consequence, a thicker stratum corneum can be formed which provides a much more effective barrier to water loss than in Amphibia. Thus, the rattlesnake loses water through the skin only $\frac{1}{40}$ as quickly as some toads and even the caiman which, for a reptile, has a fairly permeable skin, loses water $\frac{1}{2}$ to $\frac{1}{3}$ as quickly as the least permeable amphibian skin (Bentley and Schmidt-Nielsen 1966). The paucity of skin glands in reptiles helps in water conservation and the integument is always dry.

THE PATTERN OF KERATINISED EPIDERMAL SCALES

Reptiles are a diverse class because the living orders, Crocodilia, Chelonia (tortoises and turtles), Rhyncocephalia (tuataras) and Squamata (lizards and snakes), diverged early in their evolution from Amphibia. The integument therefore shows distinctive differences in the various groups. The most constant feature is the pattern of epidermal horny scales and flexible hinge regions. In lizards and snakes, the keratinised scales are rigid compact sheets of stratified flattened cells, as is also true of the larger scales in Chelonia and Crocodilia, but the smaller scales in the last two orders are composed of more loosely arranged cells, as in the hinge horny layer of lizards and snakes. The scales generally overlap each other in the Squamata which enables the skin to be stretched to a considerable extent when a snake swallows a large meal.

The horny scales are modified in certain sites to serve particular functions. Thus, the projecting abdominal scales of snakes and those under the foot pads of crocodiles provide a rough surface for gripping the ground in locomotion, while in geckos scales over the tips of the digits have become suction pads. A specialised plantar horny layer under the foot pads is never developed in reptiles as it is in birds and mammals. Sometimes the

horny scales form protective thorns or spurs as in the Australian thorny lizard.

The digits in land reptiles develop claws which are very hard keratinised structures (Biedermann 1926).

The tough reptilian horny layer provides some physical protection for the living cells underneath, but its importance in this respect is probably exaggerated since the epidermis is regularly renewed and it can withstand considerable local damage which is rapidly repaired. Probably more important in evolution was its value as a water barrier.

The epidermal scale number and pattern in reptile skin is constant for each species, which makes it useful in taxonomy, and scale differences suggest sub-speciation (Bellairs 1969). In the boa constrictor, the horny scales along the body closely correspond with the number of vertebrae, but in most reptiles this is not apparent. Each scale pattern is a reflection of the pattern of mitotic activity in the underlying epidermis, as is demonstrated by the growth rings on a tortoise carapace scale. Here the increased cell division around the edge of each scale ensures that, while each scale increases in size, the inherited pattern is not altered during growth. Clearly there is close coordination of cell division in the skin.

One environmental factor, ambient temperature, influences scale pattern, and at abnormally low temperatures snake embryos develop fewer rows of scales than in warm conditions. Lizards have evolved a protective mechanism against predators by shedding the tail, entire or in pieces. Regeneration later takes place from the stump but the new tail scale pattern differs from the original species pattern.

THE EPIDERMIS

The reptilian epidermis, excluding the keratinised layer, is usually thin, but the horny layer is often as thick or thicker than the living epidermis, both in contrast to mammals.

Proliferation is largely confined to the basal layer composed of smaller, more distinct cells than those in fish and Amphibia. Cell division occurs in cycles with intervening periods of inactivity. Generally the dermo-epidermal junction is fairly flat. The mid region of the epidermis is composed of larger prickle cells with prominent nucleoli and is rich in ribosomes for protein synthesis. These cells are all keratinocytes destined to form keratin, and glandular cells are contained in specialised glands. As these cells near the skin surface, they flatten and suddenly keratinise. Concurrently, the nucleus and other organelles are broken down by hydrolytic enzymes. Fewer cell remnants are seen in electron micrographs than in Amphibia, and reptilian horny cells rarely retain haematoxylin-stainable remnants of nuclei.

The mechanism of keratinisation and shedding of the horny cells varies in the different orders, but keratinised cells are always extremely flattened.

KERATINISATION IN CHELONIA AND CROCODILIA

In these two orders, the epidermis undergoes steady keratinisation which is directly related to epidermal proliferation. In more flexible areas such as over the neck and limbs, the horny layer appears in histological sections as a loose network of flattened squames with ruptured desmosomes over the flat surfaces of the cells. Although possibly artifactual, this is indicative of weak junctions in sites of rupture. The smaller scales in tortoises and in crocodiles show this network appearance, but the larger compact scales of the legs and carapace in tortoises and the scutes in crocodiles have the horny cells firmly stuck together. Commercial tortoise shell is the compact carapace stratum corneum of the marine hawksbill turtle *Eretmochelys imbricata.*

Cystine linkages, bound cysteine and phospholipids are distributed evenly throughout the depth of the stratum corneum in both the gopher tortoise and Nile crocodile, and not as in the uneven vertical distribution in snakes and lizards (Spearman 1969; Spearman and Riley 1969). Tortoise keratin is in the beta-pleated form as in birds, while crocodiles have mixed alpha and beta keratins (Rudall 1947). The uniform vertical distributions of both chemical constituents and of molecular types of keratin in the tortoise and crocodile indicate that keratinisation is steady as in mammals, in contrast to the cyclical changes of lizards and snakes.

Desquamation

The shedding mechanism in the loose type of horny layer is by steady desquamation of small flakes and fragments of scales visible to the naked eye readily observed in box tortoises during the summer. The loss of complete sheets is abnormal. In the compact scales the horny cells are never shed and are said to show keratinocyte retention. Year by year they pile up to form a thicker and thicker layer shown by the concentric growth rings of a tortoise carapace and larger leg scales. Each ring in temperate species represents a season's growth with the innermost areola, the horny layer at birth. In many tortoises the areola is retained for ten or more years but is eventually eroded away through wear. In consequence, the carapace scales become humped up in the middle, the oldest region of each scale. Similar piling up of horny cells accounts for the thick scales over the back of a crocodile. In a few tortoises, the carapace horny layer peels off at the end of each season's growth.

KERATINISATION AND SLOUGHING IN THE SQUAMATA

Keratinisation and subsequent sloughing of the horny layer in large sheets are similar in snakes and lizards. One factor in this process is the occurrence of cycles of keratin synthesis with different molecular forms of keratin laid down in the cells as growth proceeds. The second factor is failure of keratinisation to occur at all at one stage so that a mechanically weak fission plane is formed, although all these layers are derived from the common basal layer. The third factor is the firmness of junctions between cornified cells which are held together instead of being desquamated in small fragments. The last factor required is synchronisation of mitosis in the basal layer, necessary if the horny layer be sloughed as a continuous sheet (Spearman and Riley 1969). The keratinisation cycle complicates the histological appearance of squamate skin, which in consequence shows more morphologically distinct layers than skin with steady keratinisation. The terminology of various layers in lizard and snake epidermis was standardised by Maderson (1965). To understand the microstructure of squamate epidermis, it must be realised that layers are laid down in chronological order.

The squamate horny scales are compact structures as in the larger scales of Chelonia and Crocodilia. Indeed, the imposition of cyclical keratinisation on a horny layer showing keratinocyte retention would result in a sloughing mechanism like that in Squamata. Failure of keratinisation to occur in the nucleated cell layer is indicated by the absence of cystine bonds as well as by staining reactions which resemble unkeratinised cells. Alpha filaments present may be an unstabilised keratin precursor.

Immediately after this layer is formed, the peak of keratinisation occurs when the outermost cells of the new horny layer generation are laid down. In ultrastructure the surface has a sculptured appearance. Rudall (1947) showed by X-ray crystallography that the upper parts of the squamate scale had a beta-pleated pattern but the lower part had an alpha-helical pattern. Electron microscopy has since shown 80 Å thick filaments in the thin superficial layer. The beta layer underneath contains 30 Å filaments as in feather beta keratin. The alpha layer underneath contains 80 Å filaments as in hair alpha keratin. Histochemically, the beta region of the scale is much richer in cystine than the alpha part (Fig. 27). Clearly, there are major differences in keratins formed as growth proceeds. As disulphide bonding of the keratin decreases as each successive layer of cells becomes cornified, it is probably related to the molecular arrangement. The new unkeratinised layer is then laid down and a new horny scale is formed underneath in a second keratinisation cycle. The hinge regions show similar cyclical changes to the scales, except that the outermost cells

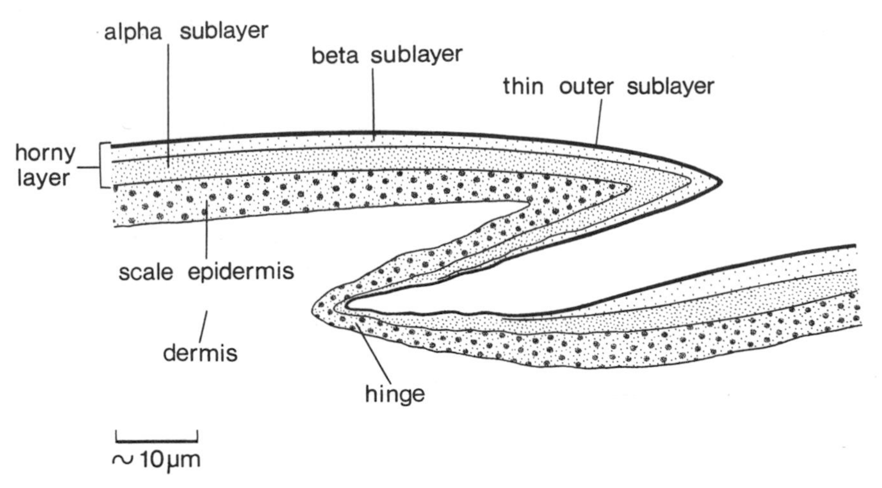

Fig. 27. The integument in a lizard, vertical section.

contain much less cystine. Prior to sloughing, eosinophilic cells, probably leucocytes, migrate out through the epidermis into the hinge regions and enter the fission plane, probably attracted by cytolysis of the dead cells.

In lizards, the old horny layer generation is usually sloughed once the horny layer underneath it is completed and is facilitated by the animal rubbing the skin against objects. Sloughing occurs in small fragments and in both lizards and snakes it includes the epithelial covering of the eyes. In a snake, the horny layer is generally sloughed in one piece. In young snakes, the old horny layer may not be sloughed when the new horny layer is formed, and occasionally as many as six generations are found in the epidermis. The squamate sloughing cycle has been divided into six stages by Maderson (1965). Sloughing occurs one or more times a year, depending on the species and on ambient temperature. It is most frequent in young animals. The first moult occurs shortly after birth or *in utero* in viviparous species such as the common adder. The rattle of the rattlesnake is made up of fragments of each horny layer generation which accumulate at the tail tip.

The horny layer in lizards and snakes contains keratin-bound phospholipids and protein-bound calcium, but not crystalline calcium salts. In the alpha part of each scale there is a surplus of protein-bound cysteine which is not all oxidised to disulphide bonds. The presence of these bound substances in squamate scales resembles the amphibian horny layer.

It appears to the author that the most important feature in the squamate sloughing mechanism is the formation of a fission plane by cytolysis of unkeratinised cells, and not breakdown of cell junctions along the interface, as occurs in Amphibia (Plate 10).

THE RHYNCHOCEPHALIA

The tuatara of New Zealand, *Sphenodon*, is the only living species in this class. These lizard-like reptiles have evolved a sloughing cycle which is superficially similar to that in the Squamata. Possibly, however, it is not the same in the tuatara since the chronological deposition of cystine in the horny layer appears to be different.

THE SKIN GLANDS

There is a paucity of skin glands in reptiles. The epidermis itself is devoid of glandular cells, but a number of small multicellular flask-shaped glands occur in certain sites in the dermis, similar in appearance to mucous glands of Amphibia which show merocrine or apocrine secretion. Others are holocrine and produce a lipoidal secretion by breakdown of cells, as in the avian preen gland and mammalian sebaceous glands. The amount of fluid secreted is insignificant and plays little part in lubricating the horny layer. Glands occur in the femoral and back skin of many lizards and under the dorsal scales of some snakes. Small glands have been described near the cloaca and under the jaw in alligators and in the dorsal skin of crocodiles (Bellairs 1969).

THE DERMIS

The integumental connective tissue in reptiles is relatively thin and there is not much stored fat. Usually the dermis is bound down rather firmly to the underlying musculature with less prominent lymph spaces than in Amphibia. There is a network of elastin fibres, and various mesodermal cells, including fibroblasts, occur, mainly in the superficial region.

Dermal ossifications occur widely in reptiles. Dermal bones on the heads of lizards and snakes are usually fused with the older membrane bones of the skull. Free dermal bones often occur in the skin, but are absent in snakes. They vary from small granules in geckos to large calcified plates and scutes in Crocodilia and Chelonia. In tortoises, bony plates occur in the carapace but have been lost in the leathery soft-shelled turtles. There are rod-shaped bones, gastralia, in the dermis of the abdomen in Crocodilia and the tuatara.

SKIN COLOURATION

Pigment cells are found in the superficial dermis of most reptiles and include large sac-like melanophores. Immediately above them in lizards there is sometimes a layer of iridophores containing guanine crystals, and

more superficially placed, a layer of lipophores. Smaller melanocytes with long dendritic processes occur in or just beneath the epidermal basal layer in many lizards, and in the pigmented spots of crocodile scales. These melanocytes transfer melanosomes to the epidermal cells, as in Amphibia, but the excess melanin is retained and they continue to function as melanophores (Spearman and Riley 1969).

RAPID COLOUR CHANGE

This is not as widespread as in Amphibia. Among lizards, it is well marked in the infra-order, Iguania, and in some geckos, but it is virtually absent in most reptiles. In those lizards which show physiological colour change, it is often marked. Chromatophore contraction and expansion is controlled in one or other of two ways. In the true chameleons (Chamoelonidae), the dermal melanophores are supplied with nerves, but in anole lizards, *Anolis*, the control mechanism is through the pituitary hormone MSH and probably adrenalin. In consequence, colour change in *Anolis* is not prevented by section of the spinal cord as it is in true chameleons.

As in lower animals, colour change is mediated through the pattern of light falling on the retina, which is readily confirmed by placing tinted spectacles over the eyes (Waring 1963). Again, the neurological mechanism is probably more complex than first appears (see chapter 14).

CUTANEOUS NERVES

These form plexuses in the dermis from which sensory nerves penetrate between the epidermal cells. Unmyelinated axons rarely occur in the epidermis and the majority of epidermal fibres have myelin sheaths. In lizard skin, the epidermal nerve fibres generally have knob-shaped terminals as in some Amphibia, and only occasionally are the ends encapsulated by epithelial receptor cells. In the dermis sensory nerve fibres either have tapered endings or they are encapsulated by groups of receptor cells: the precursors of sensory end organs of birds and mammals (Miller and Kasahara 1967). Probably these receptor cells are derived from the epidermis. Specialised sensory areas include sensory pits in certain scales and tactile prototrichs: keratinised spines which emerge from the scales of a few lizards and are supplied with nerve endings.

PHYLOGENY OF HAIRS

The suggestion that prototrichs or the sensory warts of toads might be the phylogenetic precursors of mammalian hairs seems far-fetched and unnecessary. Indeed, they have a completely different microstructure. It is

much more likely that hairs evolved in the hinge regions between reptilian scales as new epidermal appendages, for vertebrate epidermis has a general propensity to develop keratinised appendages which are not necessarily homologous, for example the breeding tubercles of fish (Spearman 1964).

BLOOD VASCULATURE

The dermis is supplied with blood capillary plexuses and lymphatics, but as the skin vasculature is not concerned in respiratory exchanges, capillaries never enter the epidermis. Vascular skin folds containing a layer of dermis form prominent crests in some lizards.

THE ORAL CAVITY

True teeth derived from fish denticles are fused to the lingual side of the jaw, *pleurodont* condition, or to the crest of the jaw bone, the *acrodont* condition, in lizards, snakes and the tuatara, but in Crocodilia the teeth are embedded in sockets, as in mammals, *thecodont*, and are not fused to the underlying bone. Succession of teeth occurs in pleurodont and thecodont reptiles, but not in those with acrodont teeth. Teeth have been lost in Chelonia which have sharp horny gums.

The tongue is an important tactile organ in snakes. In the gopher tortoise, it contains mucous cells and is unkeratinised, as in Amphibia.

SUMMARY

Reptiles were the first vertebrates to adapt fully to life on land. The skin is no longer a respiratory surface and is dry, with few glands. The epidermis develops a thick horny layer several cells in depth. The skin surface is divided into compact horny scales and flexible hinge regions. In the dorsal scales of crocodiles and larger scales of tortoises, the horny cells are retained for many years before being worn away. Elsewhere desquamation occurs in large flakes. Snakes and lizards undergo a peculiar cyclical keratinisation. At one stage the cells do not keratinise and form a mechanically weak fission plane for sloughing the old horny layer. Many reptiles have dermal bones. Colour changes are produced by melanophores controlled hormonally or occasionally by nerves, and other pigment cells occur.

10

THE SKIN OF BIRDS

ITS REPTILIAN CHARACTER

The avian skin shows many reptilian features but is characterised by a new epidermal appendage, the feather, although even this is derived from a reptilian-type epidermal scale.

The most ancient fossil bird so far discovered is *Archaeopteryx lithographica*, known from three well-preserved specimens from the Upper Jurassic period. In addition to fairly complete skeletons, there are good impressions of feathers, and even a mineralised flight feather embedded in limestone is preserved in the Munich museum. This bird, although it had plumage, had a reptilian skeletal structure. Major differences from living birds were the presence of thecodont teeth in the short beak and the long, feathered reptilian tail of some twenty vertebrae. Living birds never have teeth and even in those with prominent tail feathers, there is never a bony tail. The long bones in *Archaeopteryx* resemble those of reptiles, not hollow as in modern birds, and clearly, without the imprint of feathers, it would have been classed as a reptile. The closest reptilian ancestors of birds, from skeletal structure, were the dinosaurs, crocodiles and pterodactyls.

Birds show much less diversity than reptiles and the skins of even the most distantly related species, such as penguins and the domestic fowl, have many structural similarities (Landsborough Thomson 1964).

THE EPIDERMIS

This is generally quite thin and only two or three cells in depth, especially in sites protected by the plumage. Renewal occurs by division of cells in the basal layer. The cells pushed upwards increase in size and become prickle cells with large nuclei and nucleoli. Nearer the skin surface, the epidermal cells become flattened, and their nuclei less active. This is the transitional zone where the final stages of keratinisation occur. Above the transitional zone, there is an abrupt change to the dead horny layer made up of extremely flattened keratinised cells. In haematoxylin- and eosin-stained sections there is complete loss of DNA in the horny layer, but in paraffin sections the horny cells retain some cytoplasmic structure as in reptiles (Spearman 1971).

PLUMAGE-COVERED SITES

In sites protected by feathers, the horny layer is made up of a meshwork of flattened squames with strong lateral junctions between the cells, but with weak desmosomal attachments on the upper and lower flat surfaces which are mostly disrupted in histological sections. This appearance is very similar to the more flexible skin of crocodiles and tortoises. The relative thicknesses of the horny layer and living part of the epidermis resemble the reptilian condition.

KERATINISATION

Hydrolytic enzymes in the transitional zone of the fowl include acid phosphatase and non-specific esterase. The horny cells are demarcated from the cells underneath by their high content of cystine, cysteine and bound phospholipids. They also contain bound calcium, probably attached to the phospholipid. Fresh frozen tissue also shows free lipids mainly derived from cytolysis. Few electron microscopical observations have been made on bird skin, apart from feathers, but the light microscopical appearances suggest that the horny layer is similar to the general reptilian arrangement. In penguins the horny layer, although morphologically similar to the fowl, is very thick, which may help to waterproof the skin.

THE TARSUS

The legs of most birds are naked and the horny layer is exposed, as it is over the feet, but feathered legs occur in owls and some other birds, such as the ptarmigan. In the domestic fowl and budgerigar, the tarsus has reptilian-type horny scales and narrow hinge regions. The hinge epidermis in the fowl is morphologically and histochemically similar to that in plumage-covered skin. The horny scales sometimes contain faint haematoxylin-stainable remnants of nuclei. An important feature is the retention of strong junctions over the upper and lower flat surfaces of the keratinised cells, which makes the scales compact.

Histochemically, the main difference between the scaly tarsus and plumage-covered sites is the much stronger reaction for cystine in the scales, but with a more uniform vertical distribution than in lizard and snake scales. The molecular structure of the keratin in the avian horny layer has been less well investigated than in feathers, but it contains beta keratin, possibly with some alpha keratin (Rudall 1947).

In birds with feathered legs, feathers emerge from the apexes of the tarsal scales, as demonstrated in certain breeds of the fowl, which supports

the contention that feathers were derived from reptilian scales. Whether or not feathers ever develop from the hinge regions has not been determined.

Not all birds have a distinctive tarsal scale pattern. Water birds, such as ducks, do not show clear hinge areas, and in others the scales are very small and present a granular pattern. In the large passerine group of perching and song birds, the scaly tarsus is quite short. The scales are continued over the feet and toes in birds, a further relic of their reptilian ancestry.

THE FOOT PADS

The epidermis is thickened beneath the toes to form a tough but flexible pad for taking the weight of the bird in walking or perching, and keratinisation in the plantar region is specialised in birds as in mammals to perform this function. Development of the foot pads in birds and mammals nevertheless occurred independently, as is clearly demonstrated by the very different structure of the pad epidermis in these two classes.

The foot-pad epidermis is morphologically fairly similar in the domestic fowl, budgerigar and penguin, and so is probably constant in its essential features. It has been examined in most detail in the fowl. The prickle cell layer in the pad is thicker than in other sites, and grades into a deep transitional region. A peculiarity of the transitional cells is the way the peripheral cytoplasm and the plasma membrane stain up deeply with eosin, while the interior cytoplasm is vacuolated and only weakly stained. These cells retain haematoxylin-stainable nuclei but DNA is lost as the cells die and cornify. The appearance with eosin is reflected in the distribution of protein-bound cysteine, phospholipid, calcium, and the enzymes acid phosphatase and non-specific esterase at the cell peripheries. This is associated with the greater cystine bonding of the peripheral cytoplasm when keratinisation occurs. However, the change is more gradual in the pad epidermis than elsewhere, so that there is no sharp demarcation between transitional cells and horny layer. The latter is much thicker than in other sites which enables it to withstand the additional stresses incurred. Cell junctions remain intact in the pad horny layer so that it appears compact in histological sections (Spearman 1971).

THE CLAWS

Prominent claws are developed over the tips of the toes, particularly in birds of prey. These are made up of compact, flattened, keratinised cells held firmly together by junctional complexes. In addition to cystine bonds crystalline hydroxyapatite occurs in the cornified cells, which adds considerably to the hardness of claws (Pautard 1963).

Tarsal spurs with a bony core and covered in skin with a hard keratin sheath are present on the heels of some male birds, such as the cock pheasant.

THE BEAK OR BILL

The upper part of the beak contains a bony projection of the skull and the lower part a projection of the mandible. Covering them is a layer of skin which forms the thick keratinised rhampotheca or horny beak. The keratin layer is thick and composed of compacted, flattened cells firmly cemented together and containing hydroxyapatite, as in the claws. The hardest beaks occur in carnivores and in seed eaters, but many aquatic feeders, such as ducks, have only thinly keratinised bills and the soft skin is supplied with dermal sensory organs.

REPLACEMENT AND LOSS OF HORNY STRUCTURES

Formation and shedding of the stratum corneum and other keratinised structures in birds are closely connected with feather growth and moult, and in widely separated species the periodic moult of horny appendages has been demonstrated (Landsborough Thomson 1964). Thus, the horny claw sheath is moulted annually in the ptarmigan, as are horny excrescences around the bill of the male puffin. The outer parts of the tarsal scales are also shed at the time of plumage moult, but not in large fragments as in lizards and snakes. In the song thrush, an abnormal thickening of the tarsal horny layer was sloughed regularly at the time of feather moult, as has been found for other pathological horny growths in birds.

In the young chick of the domestic fowl, a thin zone of cells which contains little or no cystine is found in the tarsal horny scales prior to moult, and probably functions as an unkeratinised fission plane. This suggests a shedding mechanism involving cyclical keratinisation, but the horny scales are much more uniform in structure and composition than in Squamata (Spearman 1971). Clearly, much more needs to be known about the sloughing process in birds by examination of the epidermis at different stages during and between feather moults.

In penguins, neighbouring feathers are moulted together as sheets of plumage fragments. One reason for this is that the barbs on the neighbouring feather vanes become matted together, but possibly the quills may also be held together by the sloughed horny layer.

THE PREEN GLAND

The oil, preen or uropygial gland is the only important gland in bird skin. A few small glands occur in the external auditory canal and probably elsewhere, but the skin is relatively non-glandular, as in reptiles. The preen gland is a U-shaped structure visible to the naked eye situated on either side of the root of the tail with a single duct to the skin surface. The interior of the gland is made up of secretory tubules lined by germinal epithelium. The secretion is holocrine, and is rich in hydrophobic lipoidal constituents. Both the preen gland and mammalian sebaceous glands are probably derived from holocrine lipoidal secretory glands of reptiles. The oily material secreted by the preen gland waterproofs the plumage and prevents it becoming wet. It is removed by the bill during preening and is then run across the feathers; a process which also keeps the barbs and barbules correctly aligned in the vanes. Ostriches, emus and cassowaries do not have preen glands and certain genetic strains of the pigeon are without this gland with no ill effect. It is also occasionally absent in parrots and woodpeckers. Probably oiling the feathers is only essential to birds which alight on water, and in others it is not vital for survival. The main hydrophobic constituents of the preen-gland fluid are long-chain aliphatic waxes, and there are both saponifiable and non-saponifiable lipids. Glandular cells undergoing lipolysis are rich in esterases (Spearman 1971). The composition of the secretion sometimes alters during the breeding season, and presumably the preen gland is controlled by sex hormones.

THE FEATHER FOLLICLE

DEVELOPMENT

Feathers are the most complex appendages produced by the epidermis in any animal. The formation of the feather is best seen in the early development of the follicle (Lillie 1942). Plumage became important in thermal insulation when birds stabilised their body temperature and became warm blooded.

Feather germs develop as an outpushing of the epidermis and have a dermal core, in contrast to hair follicles which arise as solid downgrowths of epidermis into the dermis. At this stage, the feather follicle resembles a reptilian epidermal scale. Both hair and feather formation are closely dependent on specialised groups of dermal papilla cells beneath the follicle which induce the changes in the epidermal cells.

The rim, or collar of the feather follicle, sinks deep into the dermis and the main centre of proliferation then shifts from the basal layer of the germ to the collar matrix cells which, during feather growth, divide rapidly (Fig. 28). The epidermis of the feather germ is divided into three

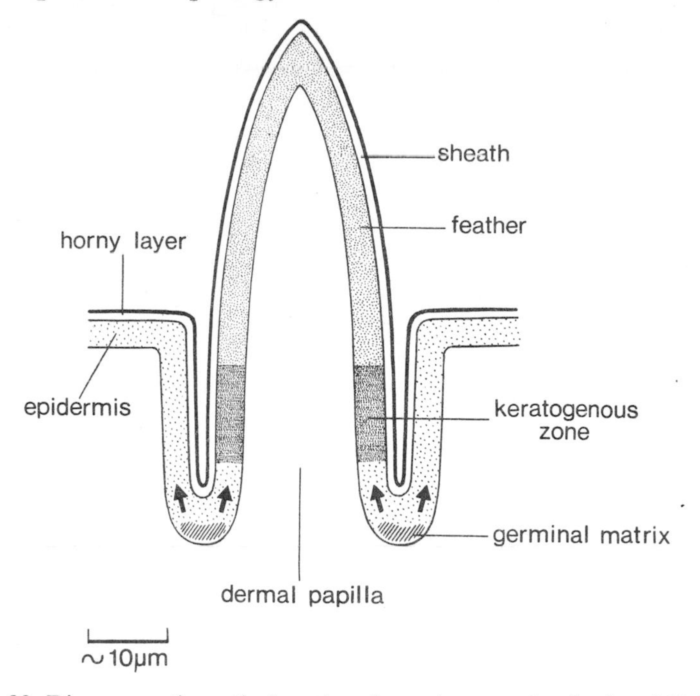

Fig. 28. Diagrammatic vertical section through a growing feather follicle.

layers: the basal layer which plays little part after the initial stages of feather germ formation, the prickle cell layer which later keratinises to form the feather rachis, barbs and barbules, and the cornified layer of the feather germ which forms the feather sheath. Surrounding the feather germ is the outer follicle horny sheath and its formative epidermis. The space between the outer sheath and the feather sheath is obliterated during growth. As the feather germ grows from the matrix, it first reaches deeper into the dermis, but then further growth pushes the germ out from the skin surface. The inter-related increase in size and elongation of feather cells prior to keratinisation is important in determining the accurate alignment of barbs and barbules in the feather vane. In flying birds, the barbules must interlock in adjacent rows which depends on good alignment, and so it is not surprising that developmental processes sometimes go wrong and abnormal feathers with non-aligned barbs result. Many genetic mutants with this type of morphogenetic defect have been described in domestic pigeons and fowls. Many flightless birds, such as the ostrich *Struthio*, have free barbules. The pattern of barbs and barbules in the feather vane results because the feather cells retain strong lateral junctions between cells in the same barb or barbules, but between adjacent structures the desmosomes break down prior to keratinisation. Cells not

destined to become part of the feather do not keratinise and are autolysed. Inward growth of the collar cells at the close of feather formation constricts the entrance to the dermal core of the follicle and the contents, deprived of blood supply, dry up to form the pith. The feather sheath then splits open and the completed feather unfurls.

DIFFERENT TYPES OF FEATHERS

The simplest feather is the down feather of the young nestling chick in which rows of barbs with barbules radiate out from a short shaft. In adult birds similar feathers often form an undercoat, but the outer contour feathers as well as the flight and tail feathers nearly always have a vane structure. This is produced because matrix cells proliferate more rapidly at one side of the collar and form a shaft, or rachis, along which the barbs are arranged (Fig. 29).

The calamus, quill, of the completed feather which surrounds the pith remains attached to the base of the follicle at the end of growth when the matrix cells enter the resting phase. Feather replacement is by renewal oı matrix activity and formation of a new feather germ immediately beneath the old feather. This differs in an important respect from hair growth in that the new hair grows up alongside the old hair (Fig. 30). Moulting in birds is therefore a mechanical process and the old feathers are only shed when the new feather sheaths to which they are attached are lost. A consequence of this is that some time elapses after loss of a feather and the unfurling of a new feather to replace it. Therefore neighbouring feathers are generally not all moulted at once, but this sometimes happens as in certain ducks which become temporarily grounded (Watson 1963).

The first moult occurs in the nestling when the down is replaced by the juvenile plumage. The latter is replaced in the autumn by the first adult winter plumage with vane feathers. Adult birds moult at least once a year; many moult twice, in spring before breeding and again in the autumn, and a few moult thrice yearly.

FEATHER AREAS (PTERYLAE)

Feather follicles are confined to well-defined areas known as pterylae. Between these regions the skin is devoid of follicles but is nevertheless covered by plumage from adjacent sites.

Feather growth usually starts at a centre in each pteryla and then spreads to neighbouring follicles with bilateral pterylae in phase.

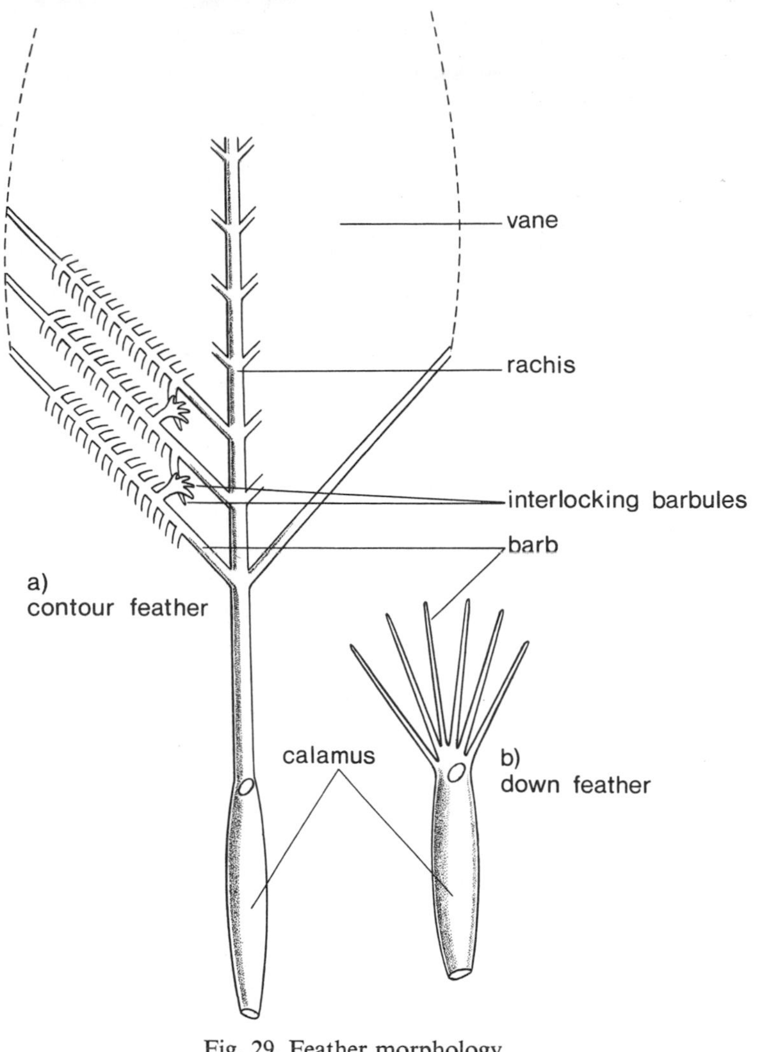

Fig. 29. Feather morphology.

FEATHER STRUCTURE

The arrangement of the calamus, rachis, barbs and interlocking barbules is shown in Fig. 29. Each feather is constructed with morphologically distinct keratin-filled cortical cells and hollow medullary cells.

FEATHER KERATINISATION

The feather-forming cells are particularly rich in bound cysteine and ribosomal RNA, some of which is carried up into the keratinised feather.

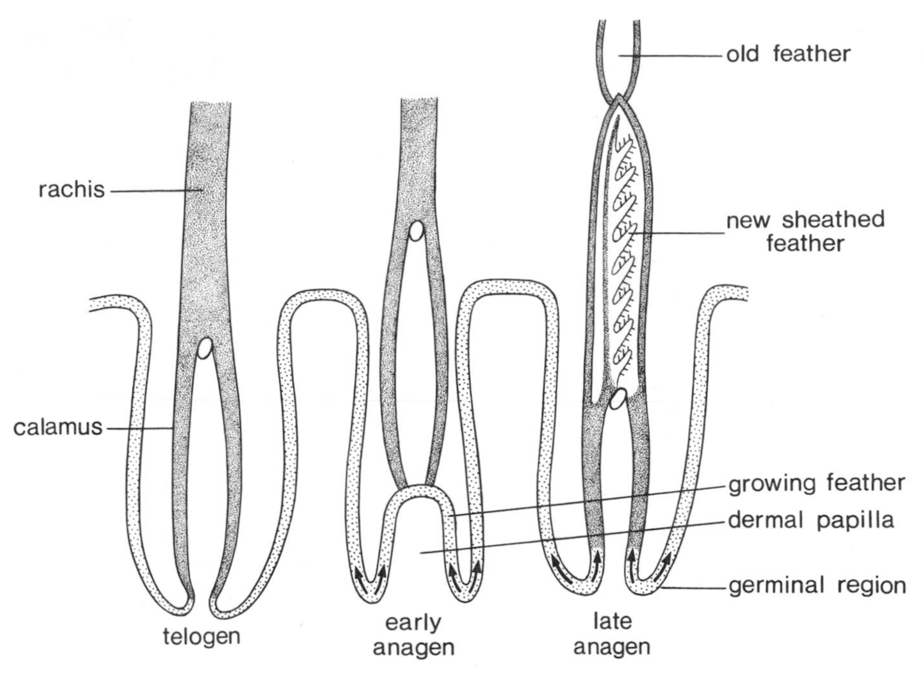

Fig. 30. Stages of feather growth and moult.

Mature feathers contain the beta-pleated form of keratin filaments embedded in a non-orientated matrix (Filshie and Rogers 1962). Both the filaments and matrix are rich in cystine bonds, in contrast to the coarser alpha keratin filaments of hairs with less cystine. Small amounts of bound phospholipid and unoxidised cystine are carried up into the feather.

PLUMAGE COLOURS

Brown and black colours are produced by melanin in melanosomes taken up by potential feather cells from melanocytes in the collar region of the follicle. Carotenoids dissolved in the keratinised cells are responsible for many red and orange colours, as in the scarlet ibis, and are dependent on diet. The bright red wing feathers in turacos are due to a copper-containing porphyrin extractable in dilute ammonia. Blue is produced by physical scattering of light on particles less than 0.6 μm across in the feather cells. Whiteness in feathers is due to reflection and refraction of incident light from innumerable surfaces presented by air in the medullary cells and surfaces of the barbules in the absence of pigment. Iridescence in humming birds is caused by light interference from the feather barbule

keratin surfaces and possibly also from melanosome surfaces. Green is produced by yellow carotenoids in combination with physical light scattering (Landsborough Thomson 1964; Ralph 1969).

THE DERMIS

The dermis is relatively thin in birds as in reptiles and contains collagen and elastin fibres, but never bone or cartilage. The superficial dermis has more cells than the deeper connective tissue. Plumage-covered skin is usually unpigmented. Blood plexuses and lymphatics occur in the dermis, and the feather follicles are especially well supplied with blood vessels. It contains pulmonary air spaces.

INNERVATION

Sensory nerves form subepidermal and deeper plexuses. Nerve endings penetrate between the basal cells, but sensory receptor cells do not occur in the epidermis except in the mouth where the tongue epithelium contains taste buds. A variety of multicellular sensory receptors occur in the dermis and are each supplied with nerves. They are similar to, but not identical with, the dermal receptors of reptiles and mammals. Unencapsulated nerve endings also occur in the skin connective tissue. Smooth arrector muscles supplied by nerves are attached to the feather follicles and fluff up the plumage in cold weather.

SUBCUTANEOUS FAT

Reserves of subcutaneous fat occur in aquatic species and are important for thermal insulation.

SUMMARY

Avian skin is reptilian in character and the only important gland is the single preen gland. Feathers are epidermal scale derivatives and undergo cycles of growth and moult. Each new feather grows up immediately beneath the old feather in its follicle. Plumage is confined to particular areas, but the vanes provide a complete covering for thermal insulation. Only the beak and toes and sometimes the tarsal skin are exposed, sites which are more strongly keratinised.

11

THE MAMMALIAN DERMIS AND SUBCUTANEOUS TISSUE

The mammalian dermis and subcutaneous tissue have been investigated more extensively than in other groups, and so require greater attention than in previous chapters. The broad problems and findings are nevertheless applicable to other animals.

CHARACTERISTICS OF THE DERMIS

FIXATION ARTIFACTS

Investigation of the dermis presents special problems because of the physico-chemical instability of collagen. Thus, reagents which promote fibre cross-binding such as formalin, used in routine histological methods, inevitably produce artifactual changes.

DIFFERENCES IN COLLAGEN AGGREGATION

The traditional microscopical appearance of connective tissue as ropes of collagen separated by ground substance clearly cannot be regarded as necessarily true in life. Tendons are made up of numerous long collagen fibrils themselves made up of microfibrils linked together laterally in bundles. There are all gradations between tendon and the teleost swim-bladder which has a fibrous gel of collagen microfibrils none of which are more than 0.3 mm in length.

ELASTIC PROPERTIES

The mechanical properties of the living skin connective tissue can be verified by the reader. Human arm skin can be readily displaced laterally over the underlying muscle and bone by the pressure of a finger. Movement is, however, limited by restraining elements, and once the finger is removed the skin returns to its original position. This shows that the connective tissue is elastic but it also has tensile strength and can be stretched only to a limited extent. Skin elasticity decreases with age due to changes in the fibre cross-bonding. Because of its elasticity, human arm skin and the skin

over the back of a cat can be pulled up into folds between the fingers and when released returns to its original state. Nevertheless, it is clearly attached to the underlying muscle layer. In a few sites, such as over the face, the palms and soles, the glans penis and clitoris, the skin is more firmly bound to the tissues underneath.

THE APPEARANCE OF EXCISED SKIN

To remove a piece of fresh skin from a recently killed guinea-pig, rabbit, rat or mouse, the dermal connective tissue has to be cut with a scalpel, and collagen fibres holding the skin to the deep fascia must also be severed. Once removed, with the epidermis face down, the dermis presents a moist gelatinous appearance. It cannot be cut to shape in most mammals but assumes a blob with rounded edges. The gel structure shows both tensile strength and elasticity. A gelatine jelly also behaves as an elastic gel but gelatine, which is heat-denatured collagen, is non-fibrous and in consequence has no tensile strength. The tensile strength of the dermis is due to its constituent network of protein fibres, and it behaves in life as a hydrated fibrous gel.

In a few mammals the dermis is different in that excised fresh specimens can be readily cut to shape. This type of dermis shows little elasticity and resembles tendon. Examples are the back skin of the hippopotamus, the skin of the walrus and probably of other marine mammals. The horse also has a tough non-elastic dermis just over the base of the tail. The belly skin of the hippopotamus which has to expand has an elastic dermis (Harkness 1968).

MICROSCOPY OF FRESH SKIN

Thin fresh-frozen cryostat sections of skin can be examined unfixed in isotonic saline by phase contrast microscopy, or they can be stained in a vital dye such as acridine orange. Guinea-pig back and human dermis, both of the elastic gel type, when examined in this way show a homogeneous birefringent mass of collagen with no visible fibres except for elastin which stains yellow with acridine orange. Distinct collagen fibres are seen in the subcutaneous tissue and are also discernible in the dermis in old age: evidence of progressive cross-bonding between microfibrils. There is much less collagen in the skin than in tendon.

Fresh specimens of the rigid hippopotamus dermis have not been examined, but its mechanical properties suggest that the collagen is considerably cross-bonded.

Collagen is aggregated from soluble tropocollagen particles: 2300 Å long and 14 Å in diameter. Free tropocollagen can be extracted from fresh dermis by cold neutral salt solutions, and is readily converted *in vitro* to

collagen microfibrils. The largest amount of soluble collagen occurs in healing wounds and in the connective tissue of foetal and young animals. In old age much less soluble collagen is present. Nevertheless, most of the collagen, even in young individuals, is insoluble, which suggests that it is present in the form of short-length collagen microfibrils. These are 1000 Å in diameter, much less than the resolution of the light microscope.

Elastin

This is always present as fine distinct fibres. It forms a network just beneath the epidermis and around the epidermal appendages with less in the deep dermis. Elastic fibres have both tensile strength and elasticity, possibly as in nylon. Elastic fibres from ligmentum nuchae (which may not be exactly the same as those in the dermis) can be stretched to twice their resting length and return to their original length. The elephant belly dermis, which has to be elastic as well as having tensile strength, is rich in elastin (Harkness 1968). The collagenous dermis of rat back skin, which contains little elastin, is irreversibly stretched rather easily during preparation of pelts for taxidermy, and even tendon collagen can be irreversibly stretched if subjected to prolonged strain.

The elasticity and tensile strength of the living dermis is due to the combined properties of collagen and elastin.

Acid mucopolysaccharides

In life, these are closely associated with the collagen and are precipitated as ground substance in fixed tissue. In foetal skin, the mucopolysaccharide present is mainly hyaluronic acid, but in adults sulphur-containing chondroitin sulphate is more common. Some glycoproteins are also present. Probably these substances influence the degree of tropocollagen aggregation. Injected labelled sodium ions diffuse readily through the dermis (Spearman and Garretts 1966), and the most dramatic demonstration of its fluid gel properties have been shown by B. Matthews in 1971, who by cine microphotography found that hookworm strongyliform larvae swim easily through excised cat and human dermis without impediment as through a viscous jelly.

The various fibrous proteins in the dermis are orientated so that they cross each other and are not parallel as in tendon, which provides the maximum mechanical strength in the skin. Harkness (1968) has pointed out that the mechanical properties of the skin connective tissue are much more complex than would appear from superficial examination.

The dermis contains up to a quarter of the body fluid, and the skin is the major organ for water storage.

THE BASAL LAMINA

This thin interfacial region of the dermis and epidermis, which occurs in all vertebrates and all but the most primitive invertebrates, has been most studied in mammals and in Amphibia. It is rich in acid mucopolysaccharides and also contains a network of fine collagen filaments and some 7 per cent of other protein (Ramachandran and Gould 1967). In tissue culture, labelled amino acids utilised in collagen synthesis are rapidly taken up by mesodermal cells and later appear in the basal lamina, and the epidermal basal layer also contributes mucopolysaccharides and collagen. The fact that the epidermis can be readily separated from the dermis in the region of the basal lamina by heating it to 60 °C suggests that the latter probably functions as a viscous adhesive with no structural attachment to the epithelium above.

LEATHER

This is tanned collagen in which the cross-linking of the fibrils is increased by treatment with reagents. Formalin can be used for this purpose and is also widely used as a histological fixative. Leather is generally made from the hides of cattle. The largest single source of supply of shoe leather is India, where the hides of dead cows are regularly shipped to Europe. The most valuable feature of leather not yet simulated by synthetic products is its ability to withstand repeated flexion without breaking (fatigue). The broad-diameter fibres in leather reflect the orientation of the collagen in the original fresh skin. This is also true of histological preparations after fixation and paraffin processing.

THE MICROANATOMY OF THE DERMIS IN FIXED TISSUE

Having mentioned problems in anatomical interpretation, the morphology of the fixed connective tissue seen in routine histological preparations will now be discussed. These show the dermis subdivided, as in other vertebrates, into superficial and deep layers.

THE SUPERFICIAL DERMIS

This contains more cells than the deep dermis and a network of elastin fibres which surrounds the hair follicles and skin glands. Growing hair follicles penetrate into the deep dermis but are always surrounded by a thin layer of superficial dermis. Collagen fibres are finer than those of the deep dermis. Surprisingly, cellulose has been detected in the superficial dermis in a number of mammals.

THE DEEP DERMIS

This is composed of broader-diameter collagen fibres and scattered elastin fibres. The so-called ground substance between the interwoven fibres reacts for acid mucopolysaccharides and also contains various cells, blood vessels, lymphatics and nerves.

THE SUBCUTANEOUS TISSUE (HYPODERMIS)

This is variable in depth but is almost always thicker than in lower vertebrates. It contains collagen fibres which bind the skin to the underlying muscle fascia and prominent lymph spaces. Relatively little fat is stored in the hypodermis in mammals from tropical and temperate regions. Examples are the rabbit, guinea-pig and common rat. Marine and Arctic mammals lay down a deep layer of subcutaneous fat (blubber), and fat occurs in human and sheep skin. The functions of subcutaneous fat are food storage and thermal insulation.

CONTROL OF FAT DEPOSITION AND LIPOLYSIS

Fat deposition is controlled by hormones, which include sex hormones and insulin. Mobilisation of fat (lipolysis) is promoted by certain prostaglandins, unsaturated fatty acid with hydroxyl groups. These are local hormones secreted by most cells of the body. Fat is transported in the blood in the form of minute globules (chylomicrons) made water-soluble by combination with protein. Adipose tissue is by no means inactive, and experiments with radio-active labels have shown continuous exchange between the blood and adipose cells with new fats deposited as others are broken down (Renold and Cahill 1965).

THE CONNECTIVE TISSUE CELLS (ROSS 1968)

FIBROCYTES

These are the most numerous cells in the dermis and are mesodermal in origin with fine dendritic processes. Fibrocytes in adult dermis are normally isolated from one another, but they form junctional complexes when closely packed together. The function of these cells is to synthesise tropocollagen, elastin, acid mucopolysaccharides and glycoproteins, although it is uncertain whether or not the same cells form all these substances. Collagen synthesis within fibrocytes has been followed by autoradiography after uptake of labelled (H^3) proline. As in exosecretory cells generally, fibrocytes have a prominent rough endoplasmic reticulum with membrane-bound ribosomes. Collagen polypeptides formed in the ribosomes pass

along the endoplasmic reticulum to the Golgi apparatus, and thence move to the exterior of the cell, probably through fine pores in the plasma membrane. Actively secreting fibroblasts in foetal skin are rich in RNA and have prominent nucleoli. In adult skin, they are much less active as only enough substances are produced to maintain the turnover in the dermis. Nevertheless, in response to wounding, fibrocytes divide rapidly and increase their RNA for connective tissue repair. Collagen synthesis is inhibited by glucocorticoid-type corticosteroids.

CHONDROCYTES

Cartilage, a solid flexible skeletal material, occurs in the pinna of the ear. It is made up of a glycoprotein-rich matrix which also contains chondroitin sulphate and a network of fine collagen and elastin. Chondrocytes secrete much more glycoprotein but less collagen and elastin than fibrocytes. They have a prominent Golgi apparatus and a rough endoplasmic reticulum, and are rich in glycogen. In adult cartilage, the chondrocytes become widely separated by their secretory product. Unlike fibrocytes, chondrocytes do not form junctional complexes. Cartilage is never penetrated by blood vessels, and movement of respiratory gases and nutrients is by slow diffusion. This limits the thickness of cartilaginous structures. Calcification does not occur in the ear cartilage although it is the first stage of endochondrial ossification in the long bone epiphyses.

OSTEOCYTES

Bone, like cartilage, occurs occasionally in the dermis, and it will be remembered that the intramembranous bones of the skull are dermal in origin. Large dermal bones are rare in mammals; the best example is the armadillo with its dorsal plates.

Osteocytes resemble fibrocytes and chondrocytes in secreting collagen, acid mucopolysaccharides and glycoproteins. They are readily identifiable in electron micrographs as the rough endoplasmic reticulum is confined to one end of the cell.

Initially in bone formation, collagen fibrils, acid mucopolysaccharides and glycoproteins are laid down. Hydroxyapatite then crystallises out on this organic material. Alignment of crystals is in consequence determined by the arrangement of collagen fibrils and continued crystallisation from ions in solution eventually fills the connective tissue. In intramembranous ossification, the first signs of bone development are the appearance of hydroxyapatite crystals between layers of osteocytes which eventually become surrounded by bone substance. The long dendritic processes of neighbouring osteocytes remain in contact through canals in the bone.

Dermal bones are at first cancellous with an irregular network of spicules and plates with blood vessels in the interstices. Compact bone is formed later by the laying down of bone lamellae by superficial osteocytes.

OSTEOCLASTS

Resorption of bone, when it needs to be remodelled, is achieved by osteoclasts, which often fuse together to form multinucleate giant cells.

BONE, DENTINE AND ENAMEL

Under pathological conditions, nodules of bone can form anywhere in the dermis, but dentine and enamel in mammals are confined to the teeth.

ODONTOCYTES

In mammals these dentine-forming cells occur in teeth. The odontocyte cytoplasm is even more polarised than in the osteocyte and has rough endoplasmic reticulum to one side and the nucleus to the other side of the cell. They are therefore different cell types. Collagen and glycoproteins are secreted which line the pulp cavity of the tooth. Hydroxyapatite then crystallises out to form dentine as in bone. A major difference from bone already mentioned in connection with fish scales is that odontocytes never get incorporated into the calcified material and only the dendritic processes of the cells become enclosed. This and the absence of blood vessels are the main differences from bone. Nerve endings do not enter dentine and sensitivity is through the odontocyte processes which may atrophy, leaving air-filled tubules. Odontocytes in human teeth and in most mammals cease to function once the tooth is completed, but in the persistent incisors of rodents, growth continues throughout life and they have to be continually ground down by gnawing (Miles 1967).

THE ENAMEL ORGAN AND TOOTH FORMATION

Enamel is the hardest and most highly mineralised material in the body. When completed it consists of a thin layer of well-orientated crystals of hydroxyapatite with only 3 per cent of organic material. The amino acid composition of the protein laid down is different to collagen and contains cystine. In tooth development, a downgrowth of the epidermis (enamel-forming organ) surrounds the dentine core of the tooth germ. The epidermal basal layer cells of the enamel organ then become ameloblasts. These exosecrete the organic material on which the hydroxyapatite crystallises out. Continued upward growth eventually causes the tooth to pierce the

epidermis. Not all mammalian teeth have the softer dentine completely covered with enamel, but exposed dentine is worn away more readily, as in the molar teeth of horses (Miles 1967).

MAST CELLS

These are normally resident in the dermis and are particularly numerous in the vicinity of blood vessels. Probably they occur in lower vertebrates, but they have been investigated mainly in mammals. Mast cells contain characteristic cytoplasmic granules, demonstrated under the light microscope by special staining methods. Under the electron microscope, the granules show a limiting membrane. They contain histamine, and in some species 5-hydroxytryptamine, released from the cells under conditions of tissue trauma, causing capillary dilation and permeability. The chief cause of this release is the antibody–antigen reaction. The blood anti-coagulant heparin is also stored in the granules. Release of these substances is followed by temporary loss of granules.

HISTIOCYTES AND MACROPHAGES

These dermal cells are able to engulf particulate matter by phagocytosis when they may fuse to form giant cells. The inactive stage is the histiocyte and the active migrant stage the macrophage. Dead tissue, particulate matter such as tattoo pigment, and foreign bodies are all phagocytosed. If the material cannot be broken down and is not toxic, it is stored indefinitely.

LANGERHANS CELLS

These mysterious dendritic cells are demonstrated only by special gold staining techniques in light microscopy and by electron microscopy, when they are found in the intercellular pathway of the epidermis. Adenosine triphosphatase occurs in human Langerhans cells. In the house mouse hairy skin, these high level cells contain non-specific esterase, and in *Potto*, alkaline phosphatase is the enzyme present. Langerhans cells do not contain tyrosinase, which distinguishes them from melanocytes. Under the electron microscope, characteristic 'tennis racquet' shaped granules are seen in the cytoplasm. The origin and functional importance of Langerhans cells are obscure. One view is that they are derived from some form of histiocyte, and dermal Langerhans cells with similar granules to epidermal Langerhans cells have been described (Breathnach 1971*a*).

MELANOCYTES (Plate 11)

These are the melanin-pigment-producing dendritic cells derived from the ectodermal neural crest which migrate into the dermis and other connective tissues during foetal life (Niebauer 1968; Riley 1967). Sometimes melanocytes remain deep in the dermis and become melanophores, as in whale skin. However, in the majority of mammals the melanocytes enter the epidermis between the basal cells. In light microscopic sections stained with haematoxylin and eosin, these cells have poorly stained cytoplasm and usually appear unpigmented with a compact nucleus, and used to be termed clear cells. By electron microscopy, the most characteristic feature is the presence in the cytoplasm of either clear premelanosomes or opaque melanosomes containing melanin. Premelanosomes are constructed of concentric lamellae, but the functional significance of these ultrastructural layers is not yet understood. In melanogenesis, tyrosinase appears in membrane-limited vesicles, the future premelanosomes, budded off from the Golgi apparatus. As the oxidation of tyrosine continues and the stages of melanin synthesis are completed, the insoluble, dark brown pigment granules appear. Melanosomes enter the melanocyte dendritic processes and are phagocytosed by neighbouring epidermal basal cells. Epidermal cells can even phagocytose carbon particles. Melanosomes in the epidermal cells are carried passively upwards to the skin surface and are lost at desquamation. In lighter pigmented human skin, the unpigmented melanocytes in the basal layer can be demonstrated by silver reduction methods or by the Dopa reaction, but in many darkly pigmented mammals, such as seals, melanin in the melanocytes can be readily seen in histological sections without special techniques. Melanocytes in the basal layer divide to provide new pigment cells, but the normal life span and fate of effete cells is uncertain.

VASCULATURE

The mammalian dermis is supplied with deep and superficial blood capillary plexuses. Anastomosing vessels enable many capillaries to be collapsed and the blood is shunted through the remaining channels. The vascular tone is determined by the autonomic nerve supply through vasoconstrictor adrenergic and probably vasodilator acetylcholinergic nerve endings. A vasodilator peptide, bradykinin, is produced by active human eccrine sweat glands. In a hunted animal, such as a gazelle, adrenalin from the adrenal medulla causes dermal vasoconstriction so that blood can be diverted to the muscles. The total skin vasculature is much greater than required for the metabolism of skin cells because the integument functions as a blood storage reservoir, the volume of which can be readily altered to

buffer changes in blood pressure. Capillaries never penetrate the epidermis and diffusion of respiratory gases, nutrients and metabolites is via the dermal tissue space and thence between the epidermal basal cells, through the epidermal intercellular pathway as far as the granular layer (Montagna and Ellis 1961).

The dermis also has a very extensive system of lymphatic capillaries which supplement the veins in draining tissue fluid back to the heart.

NERVES TO THE SKIN

In addition to the autonomic nerve supply to sweat glands, blood vessels and *arrector pili* muscles of hair follicles, the skin is richly supplied with sensory nerves which branch in the dermis to form several plexuses (Weddell, Palmer and Pallie 1955). Naked filamentous nerve endings enter the lower part of the epidermis through the intercellular pathway, but sensory receptor cells do not occur in the epidermis although taste buds are found in the tongue epithelium. Various multicellular receptor end organs encapsulate nerve endings in the skin connective tissue. The most important of these are Meissner corpuscles, which are up to 100 μm in length, are particularly numerous in the fingertips of higher primates and are primarily touch receptors, and Pacinian corpuscles which are located in the deep dermis and which are primarily pressure receptors. Numerous free nerve endings occur beneath the epidermis. Nerve networks surround the hair follicles and respond to the slightest hair movement, important in species with thick pelts and in tactile vibrissae. Sensory nerves to the skin arise from neurones in the dorsal root ganglia and brain for cranial nerves. The endings respond to heat, cold, touch, pressure, itch and pain.

Two different theories of sensation have been propounded: the older corpuscular theory which states that each sensation is distinct and depends on separate nerve endings, and the newer impulse pattern theory that a cutaneous receptor converts any stimulus into a particular pattern of impulses which are later analysed in the brain. According to the pattern theory, various types of sensation grade into one another, which is what appears to be true, and the one nerve fibre can conduct patterns for different types of sensations, a view most electrophysiologists now favour (Sinclair 1967; Iggo 1968). Nevertheless, some nerve endings appear to be preferentially stimulated by a particular stimulus and show lower thresholds of response to other stimuli. The skin is so profusely supplied with nerve endings that any point on its surface normally has one or more endings beneath it.

Impulses from these nerve endings are conveyed to the brain via the spinal nerves and sensory cranial nerves. Both slow transmitting unmyelinated nerves and fast transmitting nerves with a fatty myelin sheath

convey messages, so that in the cerebral cortex a pattern of impulses in both space and time has to be decoded to give the sensation. The number of neurones in the cerebral cortex serving the different cutaneous nerves varies from site to site, which partly explains the extreme sensitivity of the human fingertips. Sensory endings in male and female genital skin connect with a nerve centre in the sacral region of the spinal cord and trigger the sexual reflexes in addition to supplying information to the brain. If a pattern of point contacts, as in the dots of a newspaper photograph, is applied to human back skin, a picture can be visualised by blind subjects who surprisingly interpret them not in terms of touch but as a crude form of sight. This shows the versatile interpretative ability of the brain to sensory stimuli.

CUTANEOUS MUSCLES

The subcutaneous tissue of many mammals contains a thin layer of striated muscle which has its insertions in the dermis and is used to twitch the skin and drive off insects. Human facial muscles are similar. Arrector pili smooth muscles erect the hair follicles.

12

THE MAMMALIAN EPIDERMIS AND ITS APPENDAGES

The skin of most mammals is well supplied with multicellular glands which, together with hair follicles, are epidermal appendages. The remainder of the epidermis is made up entirely of keratinocytes, if one excludes the dendritic melanocytes and Langerhans cells which are not epithelial in origin. Individual glandular or receptor cells never occur in the epidermis.

PHYLOGENY OF MAMMALIAN EPIDERMIS

Hairs are peculiar to mammals and are not found in any other class. In the majority of species a thick coat (pelt) completely shields most of the body surface from view. Mammals were evolved from the extinct Therapsida, a branch of reptiles whose ancestors diverged from the primitive reptilian stock as early as the carboniferous period. Although no integumental structure is preserved in fossil specimens, pits in the outer surface of maxillary bones suggest that some late therapsids had tactile vibrissae, for similar pits occur beneath vibrissa follicles of living mammals. The first hairs were probably tactile appendages, but as body temperature became stabilised they must have soon been involved in thermal insulation. This necessitated a large increase in the number of hairs by development of secondary follicles to form a pelt. In addition to trapping still insulatory air against the skin surface, the closely arranged hairs continued to retain their sensitivity even to slight pressure, as can be verified on the human scalp. The pelt is also a barrier against physical damage.

The vertebrate epidermis shows a general tendency to develop keratinised appendages, examples of which have been mentioned in previous chapters. The evidence is against hairs being derived from horny appendages of lower vertebrates and probably they were completely new epidermal derivatives. In those mammals which have retained reptilian-like horny scales in the tail skin, hair follicles are confined to the hinge regions (Fig. 31), which suggests that hairs probably first appeared in the epidermis between the scales in certain pre-mammals (Spearman 1964, 1966).

In the scaly tails of many rodents and marsupials, groups of three or four long primary hair follicles occur in each hinge region. Elsewhere over

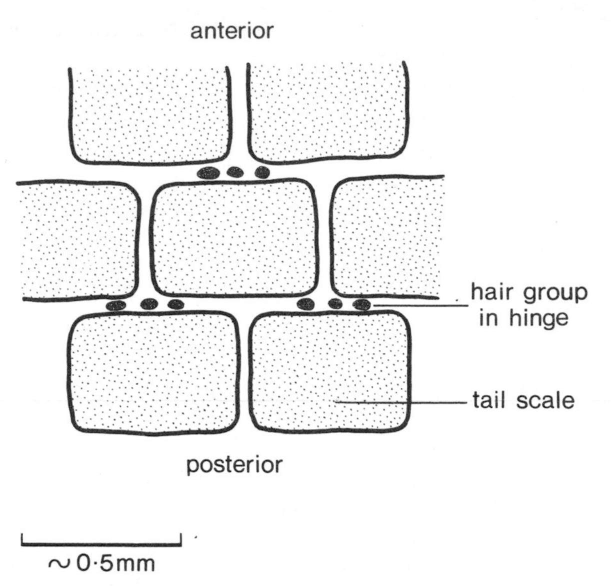

Fig. 31. Scale and hair pattern in rodent tail skin.

the pelt skin there is no trace of scales, although the arrangement of primary hairs in groups suggests that scales were originally present but were lost when the secondary hairs were evolved. Secondary hairs form the shorter undercoat or wool and are arranged around the primary hairs which form the sleek outer coat.

The appearance of large numbers of secondary hairs led to an increase in area occupied by the hinge regions and the original scales were lost. This was therefore associated with warm-bloodedness requiring secondary hair development for thermal insulation. The bare tail of the house mouse is an organ for heat loss, and here scales have been retained (Barnett 1965). In a few mammals, notably higher primates including man, primary and secondary hairs are not morphologically distinct.

The flexible horny layer typical of mammals is quite different in its microscopic structure from lower vertebrates which all have variations of the parakeratotic scale type of horny layer in both compact and more flexible regions, and the dead keratinised cells retain only partially degraded organelles. The stratum corneum of mammalian hairy skin is characterised by much more extensive autolysis of the original cell contents, which in consequence is disrupted in paraffin sections with large central spaces appearing in the horny cells. Another feature peculiar to mammals is the appearance of prominent cytoplasmic keratohyalin granules in the transitional zone immediately beneath the horny cells. These granules stain blue with haematoxylin, and in mammals it is termed

the granular layer. Keratohyalin has not been found in lower animals, except possibly in some reptiles by electron microscopy only. In mammalian skin a granular layer is confined to sites with hair follicles or sweat glands. Significantly, rodent tail scales do not have a granular layer although keratohyalin occurs around the hair follicles in the hinge regions. Indeed, the formation of keratohyalin and the associated new type of horny layer appear closely connected with hair follicle and sweat gland development in both ontogeny and phylogeny, and in human and sheep foetal skin these granules first appear in the necks of developing follicles and sweat ducts.

EPIDERMAL GROWTH AND CELL REPLACEMENT

THE CHANGE FROM BASAL TO PRICKLE CELLS

Mammalian epidermis has a morphologically more distinct basal layer than in lower animals, probably associated with frequent cell division. These cells have compact nuclei and little cytoplasm. As proliferated cells are pushed outwards towards the skin surface, they increase in size and become prickle cells. The synthetic activity of the epidermis occurs mainly in the prickle cell layer (stratum spinosum), which varies in depth from one cell in the mouse back to ten or more cells in human epidermis. In the majority of fur animals, this layer is no more than two or three cells in thickness.

In young growing animals, the area of skin surface is constantly increasing in size, achieved by lateral division of basal cells.

EPIDERMAL PATTERNS

Inherited patterning, such as in the arrangement and relative sizes of the epidermal scales in the mouse tail, does not alter as the skin area increases during growth. To achieve this pattern stability, the number of lateral cell divisions must vary from site to site according to a genetically controlled plan. However, not all skin patterns are retained in the adult. Thus, patterns of follicles are frequently altered by differential growth. The epidermis differs from site to site, and in human skin this is clearly noticeable to the naked eye after plastic surgery if skin from one part of the body is grafted into another area such as the face. Slight differences in keratinisation often occur in addition to purely morphological differences in surface patterns.

ARRANGEMENT OF CELLS IN TIERS

The majority of mammals have a fairly flat dermo-epidermal junction and, as most cell divisions occur in the basal layer, the vertically proliferated cells pushed outwards tend to be arranged in tiers (Spearman 1970*b*).

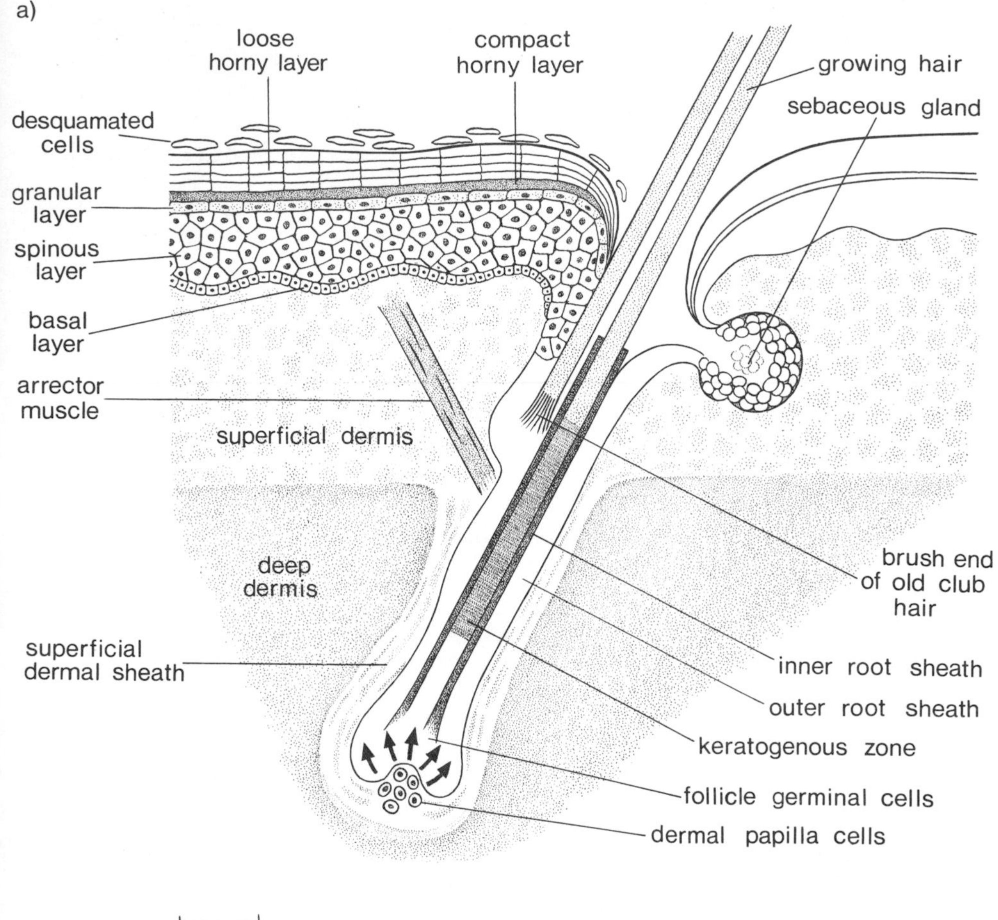

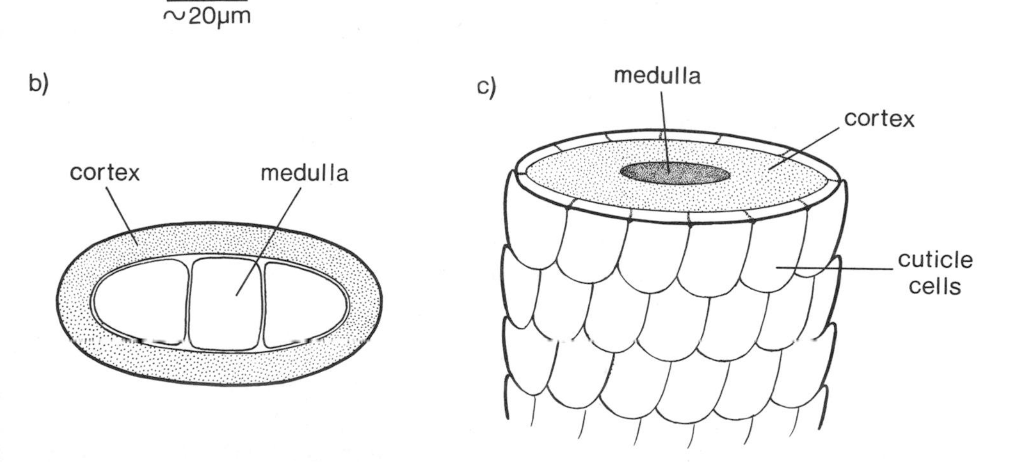

Fig. 32. (*a*) Vertical section through mammalian hairy skin. (*b*) Cross-section through hair with large medullary spaces, found in many rodents and hooved animals. (*c*) Hair fibre showing projecting cuticle cells.

Such a tier arrangement is clearly seen in the horny cells of the guinea-pig back and mouse ear, where each column of cells is proliferated from a single basal cell (Fig. 32*a*). In human stratum corneum the cell arrangement is more haphazard, probably because the dermo-epidermal junction is undulated so that the basal layer occupies a larger surface area than the horny layer.

THE MITOTIC CYCLE

During the first few days of each renewed hair growth cycle, frequency of mitosis in the surrounding epidermis is increased and afterwards subsides. Species in which single waves of hair growth pass along the body have neighbouring hair follicles in phase and related cyclical changes in epidermal mitosis are clearly shown. In contrast, human and guinea-pig skins have neighbouring follicles in different stages of activity, so that cycles of epidermal mitosis are obscured. In the house mouse, which has a single-wave growth cycle, the increase in epidermal cell division early in hair growth causes the back epidermis to become three times as thick as it was in the follicle resting phase. The peak of mitotic activity is, however, brief and as the proliferated cells become keratinised and fewer cells come up from the basal layer, the prickle cell region decreases to its original thickness. Epidermal thickness depends mainly on the depth of the prickle layer.

RATE OF KERATINISATION

For the normal balanced sequence of biochemical events leading to keratinisation to occur, the rate of cell maturation and death can be altered only within narrow limits. Therefore, the rate of proliferation is normally the most important factor in determining changes in epidermal thickness. However, the time taken for a divided cell to become keratinised varies widely in different species. Under abnormal conditions, keratinisation can occur before all the biochemical events are completed. An example is the formation of a parakeratotic horny layer with pyknotic nuclei in human skin when the normal keratin layer is stripped off by sticky tape (Pinkus 1952).

HAIRY EPIDERMIS

SITES OF PROTEIN SYNTHESIS

Above the basal layer the proliferated cells increase in size and start to form cytoplasmic structural proteins, enzymes and keratin precursor polypeptides. The prickle cells for this purpose contain numerous free ribosomes: often strung together as polyribosomes required for synthesis of long protein chains. There is a poorly developed endoplasmic reticulum,

Histochemical tests show that these cells are rich in both nucleolar and cytoplasmic RNA, and synthesis is shown by the uptake of labelled cytidine and uridine, purine bases in nucleic acids. Certain labelled amino acids are preferentially taken up into the lower part of the prickle cell layer after intradermal injection. These include leucine, methionine and phenylalanine. Others, such as histidine and glycine, are taken up by the uppermost prickle cells and lowermost granular cells, which suggests that different requirements may exist for polypeptides formed early and later in keratinisation (Fukuyama and Epstein 1969).

In adult hairy skin rapid cell death occurs in the uppermost granular cells, so that the dead horny layer is sharply defined. In this type of epidermis both mitochondrial dehydrogenases and non-mitochondrial pentose pathway dehydrogenases are absent from the uppermost granular cells which appear semi-moribund with permeable plasma membranes (Jarrett 1973). The prickle cell layer utilises mitochondrial enzymes for tissue respiration, and evidently cell death is rapid as the transitional stage is reached.

MAMMALIAN KERATIN

Biochemical analysis of homogenates of unkeratinised epidermal cells shows a mixture of proteins which must include keratin precursor polypeptides.

Keratin is a high molecular weight protein with a complex structure, built up in stages. First, different polypeptides of relatively low molecular weight, coded by different genes, are elaborated on the polyribosomes in the prickle cells, and these are then linked together by the mechanical process of complementation. Mammalian keratin from analysis of horns, hairs and hooves, is formed by the linking together of three different kinds of sub-units, each a product of complementation from a variety of polypeptides. There is a cystine-poor, mainly alpha-helical fibre component, a cystine-rich, globular matrix, and a cystine-poor, tyrosine- and glycine-rich globular matrix. Gillespie (1965) suggests that differences in keratin composition in various sites of the same animal are probably due to variation in proportions of these sub-components. Mammals differ from birds in the presence of alpha-keratin and absence of beta-keratin.

In the epidermis these final stages of keratin complementation appear to occur in the uppermost living cells and do not require vital processes. The small amount of keratin in the mammalian horny layer precludes reliable chemical analysis, and greater reliance must be placed on histochemistry and autoradiography.

HISTOLOGICAL CHANGES ASSOCIATED WITH KERATINISATION

The pre-keratinised cells

Tufts of tonofilaments in the epidermal cells, as in all epithelia, radiate from each desmosome into the surrounding cytoplasm and may curve back to neighbouring desmosomes. These are either restraining, or contractile, fibrils. In addition, the products of keratin synthesis are presumably seen in electron micrographs but cannot be identified in the absence of labels.

The weakness of speculation on the functional significance of ultrastructural features based entirely on morphological comparisons is obvious, but some elaborate theories of epidermal keratinisation have been built up on this insecure basis (see Brody 1962; Zelickson 1967; Breathnach 1971*a*). In the last decade, extensive information on the localisation of epidermal chemical constituents has accumulated at the light microscopical level, but this is only just beginning to be related to ultrastructural findings through new histochemical methods.

Light microscopical histochemistry shows a thin intense band of protein-bound cysteine SH diffuse in the granular layer. Above this level, rapid exergonic oxidation to cystine occurs with formation of keratin. In hairy skin, which includes human epidermis with its short hairs, the keratinised cells do not contain excess bound SH which is either oxidised to SS bonds or is autolysed.

Hydrolytic enzymes formed on ribosomes in the prickle cells are carried up into the granular layer either in an inactive form or contained in lipoproteinous bodies; lysosomes which vary in size are generally smaller than mitochondria (Jarrett 1973). When released into the granular cells these hydrolases break down most of the original cell structure. This autolysis parallels the final stages of keratin deposition (Spearman 1966). Once keratin is formed in the cell it is protected from autolysis by its disulphide bonds, so that finally nothing but the thickened plasma membrane, some lining cytoplasm, and keratin filaments remain in the cell. Everything else is degraded to fluid products. Acid phosphatase released in the granular layer is probably concerned in the dephosphorylation of nucleotides, liberated in the breakdown of nucleic acids (Jarrett 1973). In the outermost granular cells, the nuclei are shrunken and are surrounded by a zone of cytoplasmic lysis. Only very rarely is the remnant of a nucleus seen in a newly cornified cell.

The final stage of keratinisation is extremely rapid in hairy epidermis and involves the complete destruction of the nucleus together with other cell organelles, even including their retaining membranes. In no other type

of keratinised cell does such a complete breakdown of cell structure occur. It is this which is so characteristic of mammalian epidermis and distinguishes it from all other types.

Keratohyalin

By light microscopy these granules vary in size and staining reactions in different species and even in different sites of the same individual. This is because keratohyalin is not a definite substance but a complex of calcium, phospholipid and protein. Probably granules vary slightly in composition from site to site. The dark blue staining reaction with haematoxylin is due to mordanting with calcium and other metals present, also shown by micro-incineration. The phospholipids are probably derived from degraded cell membrane lipids and can be extracted in pyridine, a specific solvent, except after prolonged protein fixation, which suggests linkage to protein.

The significance of the protein component of keratohyalin has occupied the attention of many electron microscopists during the last decade. Chemical analysis of isolated centrifuged keratohyalin proteins has demonstrated various amino acids (Matoltsy and Matoltsy 1970). It has been suggested that some components of keratin may be derived from the various keratohyalin proteins. Nevertheless, the undisputed fact that in most forms of keratinisation keratohyalin does not occur, shows that it is not essential for keratin formation. Both phospholipid- and sulphur-rich protein stains with osmium used as a fixative in electron microscopy, which makes interpretation of keratohyalin morphology difficult. However, after partial extraction in lipid solvents interpolated between brief glutaraldehyde protein fixation and osmium post-fixation, the granules lose some of their density, appear speckled and begin to break up into osmophilic probably protein fragments, separated by spaces which contained the phospholipid. It is these fragments which some think are a keratin component.

THE STRATUM CORNEUM IN HAIRY SITES

In hairy skin this is made up of a thin basal compact region adjacent to the granular layer which contains keratin-bound phospholipid and a less compact, much thicker region above it without bound phospholipids (Fig. 32*a*). Loss of phospholipids is due to continuing autolysis within the dead cells as they are pushed upwards. The lowermost keratinised cells, having lost most of their cytoplasm, together with water by evaporation, are extremely flattened. Phospholipid in this layer may be derived from keratohyalin, but it also occurs in parakeratosis without a granular layer. The basal region of the horny layer in terrestrial mammals is important in

passively restricting water movement through the skin. In electron microscope sections, randomly arranged filaments in the keratinised cells are resistant to autolysis, and so may be the same as alpha microfilaments of hair keratin. As in the granular layer, observation of filaments in the basal part of the horny layer is obscured by phospholipid, but interpretation is greatly facilitated by partial extraction in pyridine prior to osmium. The filaments are most clearly seen in the outer horny cells in which phospholipids have been broken down. Cystine-rich matrix keratin is localised in the periphery of each cell (Spearman 1970*b*; Fukuyama and Epstein 1969), in the thickened plasma membrane and thin shell of lining cytoplasm (Farbman 1966). The interiors of the mammalian epidermal horny cells lack the stabilising matrix keratin of parakeratotic cells of Amphibia, although interestingly even there cystine is concentrated in the cell peripheries, the site of highest oxygen tension (Figs. 33*a* and *b*). Possibly when the internal cytoplasm in horny layer cells is stabilised by keratin matrix protein, it is linked by glycine and tyrosine residues. After solvent extraction and osmium post-fixation of guinea-pig hairy skin, a thin shell of peripheral cytoplasm remains in the horny cells, but the interiors show only a filamentous network surrounded by empty spaces (Plate 12). Previously electron microscopists believed that these spaces contained matrix keratin as in the hair cortex, but this was due to confusion with lipids precipitated by osmium. The rapid penetration of paraffin wax into these autolysed interiors of horny cells in hairy skin disrupts the fibrillar network and results in the characteristic central cavities seen by light microscopy (Spearman 1970*b*). Frozen cryostat sections show some stainable material in the cells since this is not dissolved out. The mechanical strength of these horny cells clearly resides in the peripheral keratinised cytoplasm and thickened plasma membranes, and the internal filamentous network is probably of little mechanical significance. This type of cornified cell therefore shows spatial separation of the matrix and fibre keratin components. Cystine-bonded plasma membranes are rare in keratinised cells and may be confined to the mammalian stratum corneum.

DESQUAMATION IN HAIRY SITES

A continuous intercellular pathway exists from the dermis to the granular layer where fused junctions occur, but in the horny layer the cells have lateral flange-shaped junctions which do not readily allow dissolved substances to pass, although they do not have fused membranes. As the horny cells move outward towards the skin surface, junctions begin to break down above the compact lipid-rich layer, probably by enzymic hydrolysis. Characteristic small cytoplasmic lysosomal bodies (membrane coating granules, Odland bodies), previously mentioned in Amphibia, occur in the

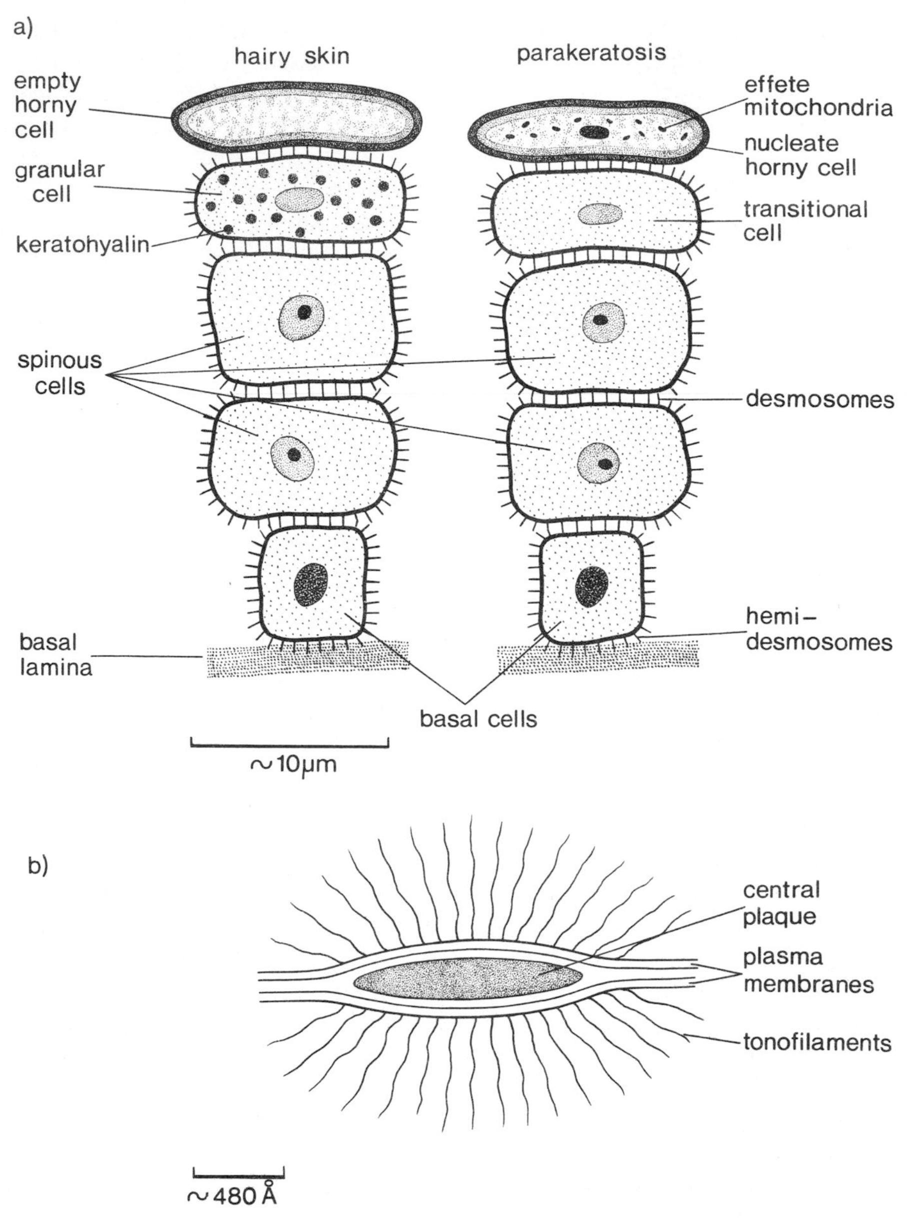

Fig. 33. (*a*) Mammalian hairy skin with highly autolysed horny layer cells compared with less autolysed cells in parakeratosis. The horny cell plasma membranes in hairy skin are more resistant to damage by reagents. (*b*) Desmosome (high magnification).

granular cells. These bodies pass out through the plasma membrane and accumulate in the intercellular spaces. The enzymes released may break down the cement between cell junctions, which causes separation into individual cells which are then lost (Rowden 1968). Brody (1966) also found both fibrous and amorphous material in the intercellular spaces of the horny layer, possibly the last remnant of the epidermal cuticle. Desquamation of individual horny cells occurs continuously over hairy skin, but in species with a covering pelt they are trapped until the hairs are moulted. In the human scalp, it is called dandruff or scurf.

The above picture of cornification probably represents in broad principles the process which occurs in hairy epidermis of most terrestrial mammals.

PLANTAR EPIDERMIS

The soles never develop hair follicles, although sometimes hairy epidermis extends partly over the under-surface, as in polar bears. The true soles contain eccrine sweat glands, generally absent elsewhere. In a wide variety of mammals from man and the guinea-pig to the tiger, the histology of the plantar epidermis appears remarkably similar in paraffin-processed sections. There is always a thick granular layer, absent in avian plantar epidermis, and above the granular layer is a thin hyaline layer, the stratum lucidum. The plantar horny cells do not have stainable nuclei, but are less completely autolysed than in hairy sites and they retain some cytoplasmic structure with bound phospholipid and cysteine in the interiors of the cells, indicative of matrix keratin. Nevertheless, the peripheral zone in each horny cell contains most cystine. Intracellular spaces are not seen in paraffin sections.

In ultrastructure, the interiors of the cells contain keratin filaments and also degraded organelles not found in hairy sites, but no empty spaces. An interesting feature found in the cavy pad, and which may occur generally, is the method of joining of neighbouring cells which are interdigitated by long cytoplasmic processes. This provides strength against shearing forces.

In the New World monkeys, the bare clasping region of the prehensile tail closely resembles a palm, and it has a rather similar epidermal histology under the light microscope. The granular cells in guinea-pig plantar skin contain pentose pathway dehydrogenases, and cell death is more protracted than in hairy skin. This respiratory pathway can still be used while mitochondria are being broken down (Jarrett 1973).

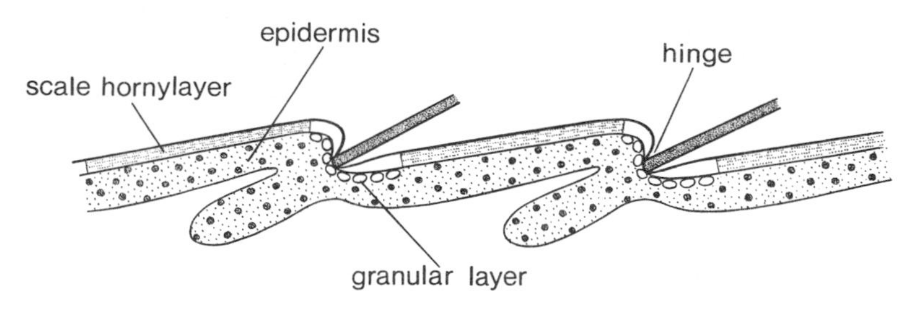

Fig. 34. Vertical section through rodent tail epidermis.

MAMMALIAN TAIL SCALES

The keratinisation of mouse, rat and marsupial tail scales is basically similar to that of reptiles and birds, and a granular layer is not developed (Fig. 34). Mammalian tail scales occasionally show faint haematoxylin-stainable remnants of nuclei and, as in birds, there is very limited autolysis of the cytoplasm which contains peripheral cystine-rich keratin with some unoxidised cysteine. In contrast to lizards and snakes which have cycles of keratinisation, keratin disulphide bonding is uniform throughout the scale thickness in mammals, which is to be expected with steady cornification (Spearman 1966). Bound phospholipid also occurs through the tail scale horny layer thickness. The transitional cells contain pentose pathway dehydrogenases as in the pads.

PARAKERATOSIS

In this form of keratinisation, the horny layer retains prominent, haematoxylin-stainable pyknotic nuclei with only partly degraded DNA. Intracellular spaces are not seen in paraffin sections, and there are effete remnants of the cell organelles. A variety of filaments and intervening nonfibrillar material can be seen (Brody 1962), but there is no granular layer (Fig. 33). In hairy mammals parakeratosis occurs normally over the moist nose in some species, as well as in places in the oral cavity and possibly in the vagina. Parakeratosis is normal in whale skin. It also occurs in certain skin disorders of man and animals and, as mentioned earlier, in an emergency if the normal horny layer is suddenly removed. Parakeratotic cells, like the tail scale cells, contain bound phospholipids and cysteine as well as peripheral cystine. There is therefore a close resemblance between parakeratosis and the tail scale type of keratinisation of some mammals

and with parakeratosis in Amphibia. In mammalian parakeratosis, cell death occurs gradually, as shown by pentose pathway dehydrogenases in the transitional zone.

NORMAL HYPERKERATOSIS

Seals and sea lions have a thick horny layer which resembles the scale type except that a granular layer is developed. These features make it hyperkeratotic, a term derived from pathology (Spearman 1970*b*). The phospholipid-rich horny layer may be useful in water-proofing the skin. In the large unrelated tropical pachydermatous mammals: elephant, rhinoceros and hippopotamus, the thick skin also undergoes hyperkeratosis (Spearman 1970*a*).

CONTIGUOUS ECTODERMAL EPITHELIA

The oral cavity is lined with epithelium which in places is clearly keratinised, but not elsewhere. The most strongly keratinised region is the tongue; in particular, the filiform papillae, which in species with rough tongues such as the cat are discrete appendages with cystine-rich spines, used to rasp food, lap water droplets and comb the coat. Less strongly keratinised filiform papillae occur in rodents, but human tongue papillae are much less keratinised.

The external auditory meatus has a keratinised epithelium, but the epithelial surface of the eye, the transparent non-glandular cornea and surrounding mucous epithelium, the conjunctiva, keratinise only in vitamin A deficiency. The vaginal epithelium undergoes cycles of keratinised and mucous phases controlled by hormones. Thus, in laboratory rats, oestrogens induce cornification. The keratinisation process, however, has not been examined adequately and shows species differences. Cells are examined by the smear technique.

Individual mucous cells in vaginal epithelium are possibly innervated as in cyclostome skin, since secretion of watery fluid occurs immediately following sexual arousal. There are no compound glands in the mammalian vagina to explain this.

SITE-TO-SITE VARIATION

Differences in epidermal keratinisation over the palms and soles and hairy sites, and regional differences in skin texture have been mentioned. Less obvious differences in the horny layer as well as in epidermal thickness and morphology also occur from region to region. Examples are the ridge pattern in the lower surface of human epidermis which differs from

site to site, and the much deeper ridges in the snout epidermis of many mammals. In higher primates, ridges with sweat duct openings on the exposed outer surface of the palmar epidermis help in gripping objects by moistening the surface and are responsible for fingerprints.

TYPES OF MAMMALIAN EPIDERMIS

Four main types of hairy epidermis occur:

(*a*) The majority of species with a covering pelt have a thin epidermis only two or three cells in depth. Examples are the sheep and house mouse.

(*b*) Human epidermis is peculiar in that it is much thicker than in most mammals, even than in the gorilla, but the horny layer is thin. A possible reason for this is the absence of a protective pelt.

(*c*) The elephant, rhinoceros and hippopotamus also have an exposed epidermis but with a thick hyperkeratotic horny layer (Spearman 1970*a*).

(*d*) In aquatic mammals, Pinnipedia have a reduced number of hairs, Sirenia have very few hairs, and whales are without hair follicles except for one or two on the head. These aquatic mammals have hyperkeratosis or parakeratosis.

SHEDDING OF THE HORNY LAYER IN LARGER PIECES

Desquamation after parakeratosis and hyperkeratosis occurs in much larger flakes than in hairy sites which shed individual cells. This is due to retention of firm junctions in the horny layer. In the elephant seal *Mirounga* and in a few other Pinnipedia, the horny layer is sloughed once a year as an intact sheet fused to the moulted hairs (Spearman 1968*a*). However, despite a superficial resemblance to snakes, the shedding process is quite different. Most other seals shed their horny cells in flakes. Desquamation in the mammalian plantar horny layer is probably as individual cells, as in hairy skin, since the interdigitations separate in the cells about to be shed.

HAIRS

Hairs are constructed from orderly arranged columns of keratinised cells proliferated upwards from localised downgrowths of the epidermis, the hair follicles. They are therefore formed differently from feathers, but are also organised structures which grow to a maximum length and are moulted as a whole, as in the snake and lizard stratum corneum.

GROWTH AND KERATINISATION OF THE HAIR FIBRE

Hairs vary in size from fine Merino sheep's wool to the coarse spines of the porcupine and echidna. Each hair is covered with flattened, imbricated cuticle cells, which are particularly rich in disulphide-bonded matrix keratin, but do not contain alpha-helical filaments (Fig. 32*c*). Over the outer surface of the cuticle cells is a thin amorphous, non-keratinous, but chemically resistant epicuticle which is not readily wettable. The cortex is made up of spindle-shaped cells which contain only slightly less cystine than the cuticle cells and are packed with alpha-keratin filaments orientated along the length of the hair axis. Although hair cells retain membranous remnants of nuclei, DNA is lost during keratinisation. At the centre of the hair there is generally a soft cystine-free medulla. Rodents and many hooved animals, such as llamas and antelopes, have large intracellular spaces within the medulla (Fig. 32*b*). Porcupine quills have a large medulla honeycombed with cavities, and most coarse hairs contain medullary spaces. The arrangement of a cortical cylinder around a central cavity gives mechanical strength and permits large hairs to be bent without damage. Many fine hairs have no medulla. Examples are Merino wool and human velus body hairs. Nevertheless, mouse pellage hairs, which are very fine, have medullary spaces, but the coarser vibrissae have a wide cortex and narrow medulla with few spaces, as in human scalp hairs.

The germinal cells of the follicle bulb divide during the growing *anagen* growth stage. Initially, cell proliferation causes the growing follicle to push deep into the dermis, but later cells are pushed outward towards the skin surface (Fig. 35). The outermost ring of germinal cells in the bulb is destined to form the outer root sheath, which is virtually a continuation of the epidermal prickle cell layer. Further in are the inner root sheath germinal cells, and centrally the hair formative cells (see Montagna and Parakkal 1973; Chase and Silvers 1969). The inner root sheath surrounding the growing hair is formed partly from the bulb, but higher up it also receives cells from the outer sheath. It grows upwards with the hair but breaks down by autolysis and is lost below the sebaceous gland opening. Remnants of this sheath are often still attached to emergent hairs. The upper part of the inner sheath stains pink with eosin, and may be keratinised. It contains bound phospholipid, probably derived from large cytoplasmic trichohyalin granules in the inner sheath nearer the bulb, quite different from keratohyalin, which does not occur in hair follicle. Sometimes haematoxylin-stainable nuclei occur in the inner root sheath.

Hair-forming cells become elongated and spindle-shaped as they enter the broad keratogenous (transitional) zone which is rich in both bound cysteine and phospholipid. The keratogenous zone contains non-mitochondrial pentose pathway dehydrogenases, indicative of slow death

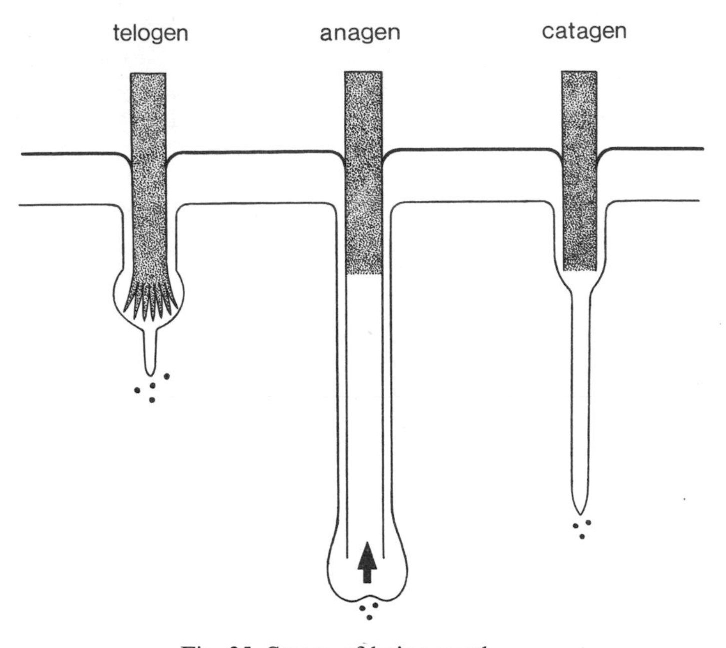

Fig. 35. Stages of hair growth.

(Jarrett 1973). There is less RNA than in feather formative cells except in the inner sheath. Isotope experiments suggest that cysteine may enter the hair keratogenous zone by lateral diffusion through the root sheaths (Ryder and Stephenson 1968). Alpha pre-keratin first appears in the differentiating cortical cells, while autolysis in the keratogenous zone removes most of the cell structure except for nucleate membranous remnants (Spearman 1966).

Below the sebaceous gland opening, the hair cells become disulphide-bonded by exergonic oxidation of protein-bound cysteine to cystine, with release of energy. At this time there is considerable shrinkage in hair diameter due to loss of water. The keratinised spindle-shaped cortical cells contain alpha-keratin microfilaments bonded together by sulphur-rich matrix keratin. The fully keratinised hairs contain practically no trace of cysteine or phospholipid and no nucleic acids. The cells are therefore virtually keratin-filled shells. Neighbouring hair cells are cemented together by much stronger junctions than in the surrounding stratum corneum, but these can be separated by prolonged digestion in proteases. Keratin filaments do not pass between cells. The parallel orientation of hair keratin microfilaments enables their molecular structure to be examined more readily than in the horny layer. Transverse sections of sheep's wool cortical cells in high-resolution electron micrographs show that each 80 Å diameter

microfilament is composed of eleven 20 Å protofilaments arranged as a ring of nine surrounding two others (like cilia). Two or three alpha-helical keratin molecules spiral around one another to form each protofilament, while each of these is in turn super-spiralled to give the microfilament, as in a length of rope (Fraser 1969). Dimensions of microfilaments do not vary much in different mammalian hairs. These alpha-helical filaments contain only about one-third as much cystine as the globular matrix and they are mainly stabilised by hydrogen bonds, a linkage between a hydrogen atom and an atom of oxygen or nitrogen. Hairs contain much more cystine-rich matrix than stratum corneum but no appreciable tyrosine- and glycine-rich matrix keratin. The arrangement of microfilaments in sulphur-rich matrix has been likened to the steel rods in reinforced concrete, and confers mechanical strength.

Once the maximum length of the hair is reached, the bulb cells cease to divide. The lower part of the follicle, including the inner sheath, is then autolysed with the exception of the germinal cells which remain connected with the upper permanent part of the follicle by a thin cord of shrivelled cells. Soon this shrinks upwards so that only the permanent part of the follicle and the germinal cells remain. This breakdown process, which is completed within twenty-four hours, is the *catagen* stage. The completed club hair remains firmly attached to the resting *telogen*-stage follicle by brush-like cystine-rich keratinised cells. When hair growth is renewed, the new hair grows up alongside the old hair, in contrast to feathers which grow up underneath the old feathers. This is much more efficient and ensures that hair moult occurs without loss of the pelt. Moult is not a simple mechanical process as in feather loss, but is dependent on as yet uncertain physiological changes in the follicle. When anagen recommences, the germinal cells divide again and the transient part of the follicle below the sebaceous gland is reformed.

In most mammals pelt hairs can be raised up by contraction of (arrector pili) smooth muscles attached to the follicles. In cold weather this increases the amount of warm air trapped beneath the hairs.

HAIR DISTRIBUTION AND ANGLE OF EMERGENCE

Hairs usually emerge from the skin at less than a right angle which ensures that the overhairs form a sleek surface to the coat. Fibres in different sites point in different directions. Over the head, back, flanks and underside of the body the hairs usually point towards the tail. Hair follicles are not confined to particular areas as are feathers, but bare areas occur, such as behind the cavy ear covered by fur from neighbouring sites. It is advantageous for sensory vibrissae to stick out from the skin surface, and so these follicles are always surrounded by an erectile blood sinus.

HAIR CYCLES

Some mammals are born naked and it is some days before the hairs erupt. Often several pelages grow and are replaced before the adult coat is formed. Sometimes, as in the juvenile cream-coloured, wavy coat of the young seal and the striped coat of the young wild boar, differences between the coats are marked.

In most wild mammals single waves of hair growth with neighbouring follicles in phase move along the body from one region to another: dorsal to ventral, head to tail, or vice versa. In consequence, at a particular time some sites may have only anagen follicles, others telogen follicles, and others are in catagen (Ebling 1965; Ling 1970). Wild mammals generally moult the old club hairs soon after the new hairs have been completed. Retention of club hairs in succeeding hair generations so that more than one old hair remains in the follicle is more common in domestic breeds, such as the dog. The majority of wild species grow a new coat and moult the old one over a short period, either once or twice yearly in autumn and in spring. When two coats are produced each year, these may differ in length and in colour, as in the Arctic fox which does not produce melanin in the winter. Hair cycles in rats and mice occur much more frequently. The wild moufflon and domestic Wiltshire shorthorn sheep moult each year, but the Merino breed has a very long anagen covering many years.

A patchy moult with no distinct waves occurs in camels, while in man and the guinea-pig neighbouring hairs are not in the same stage of growth, probably because different types of hairs have different growth cycles which are out of phase. In the normal adult human scalp some 13 per cent of the follicles are in telogen at any one time with a range of 4–24 per cent in adult Caucasian males. Anagen lasts for a few years in man, but the telogen period is short. The final length of a hair is determined both by the rate of hair formation and by the duration of growth.

HAIR PIGMENTATION

Melanocytes in the follicle bulb transfer melanosomes to formative cells of the hair medulla and cortex. Colour patterns are produced by chemical differences in the type of melanin deposited during hair growth. In many wild rodents, the tips of the hairs contain yellowish melanin, but the lower parts of the fibres have a dark brown melanin which results in a speckled (agouti) appearance, after the rodent of that name. In the porcupine quill, bands of light and dark melanin alternate. Another type of colour patterning is seen in the leopard spots or zebra stripes, in which dark and light pigmented areas occur. Melanin in light or dark forms is the only important pigment in the mammalian coat. The green fur in certain

monkeys, *Cercopithecus*, is caused by yellow melanin combined with blue produced by light interference. The distribution of different types of melanin is determined by similar complex genetic mechanisms in different mammals, described by Searle (1968).

COMMERCIAL USES OF HAIR

WOOL

Considerable information on the chemistry and molecular structure of keratin has come from wool research (Ryder and Stephenson 1968). Wild sheep have both long (primary) overhairs with a medulla, and an undercoat of (secondary) non-medullated wool. Overhairs occur in some domestic breeds but not in others. Domestic sheep have a wave cycle with a long anagen period alternating with a very brief telogen period, as in the human scalp: the reverse of most wild mammals which have prolonged resting periods. Sheep also have a much greater density of secondary follicles in the skin than other mammals. This is due to an increase in the numbers of secondary follicles by budding from other follicles. The finest-diameter wool fibres are produced by the Merino breed which originated in Spain and now is kept mainly in the chief wool-producing countries of Australia, New Zealand and South Africa. Flocks of Merino sheep were once the mainstay of the Australian economy, and although the post-war boom was ended by the upsurge of artificial fibres, improved fabrics can now be manufactured using wool and synthetics with the advantages of both. Raw wool contains sebaceous lipids, 'wool fat' (lanolin), which is extracted before preparation of yarns. British hill sheep produce a short wool useful in carpets and tweeds, but the Merino does not do well in our wet climate.

The maximum length of the shorn fleece (staple length) varies in different breeds from 1 to 18 inches. Two types of yarns can be manufactured. Woollen yarns have very short fibre lengths intermingled at different angles so that the ends project to give a bulky, fuzzy thread, useful in blankets and tweeds. Worsted yarns for knitted garments contain longer staples with twisted fibres arranged parallel with the length of the yarn, and in consequence the threads are finer and smoother. The best knitting yarn comes from the Australian Merino (Botany wool).

A useful feature of wool is its ability to absorb a considerable amount of moisture (up to 18 per cent) without feeling damp. Another valuable feature is its natural waviness (crimp) with the property, shown well in Merino wool, of neighbouring fibre waves being in phase. This gives a springy, soft texture to woollen cloth. A feature which has both advantages and disadvantages is the projection of the cuticle cell edges in sheep's wool which gives a rough feel and causes neighbouring fibres to bind

together. It also enables yarns to be spun and is the basis of felt manufacture. A disadvantage is shrinkage caused by the backward creep of one fibre over another during washing, so that individual fibres become looped and curled up.

Carpet beetle and clothes moth larvae produce an enzyme which breaks down the protective disulphide bonds in keratin, which can then be digested by proteases in the gut of the insect.

A few other mammals have been domesticated for their wool, but none has achieved the economic importance of the sheep. The angora goat, which produces mohair, and the angora rabbit are the main examples.

FUR ANIMALS

Certain species are bred commercially in fur farms for the fine quality of their pelts. Important farmed fur animals are mink, chinchilla and musk rat. The animals are killed immediately after the autumn moult when the glossy winter coat with its new melanin has just entered telogen (prime pelts). Lamb skins are also widely used for furs. Many wild fur animals have been exploited almost to extinction by hunting and trapping, and a few are protected species no longer accepted at the main fur auctions.

MISCELLANEOUS APPENDAGES

PAIRED HORNS

Horns which develop on the frontal bones of cattle, goats, sheep and antelopes have a bony core. They are covered with a non-hairy epidermis which keratinises to form a dense horny layer, never shed except in the American pronghorn antelope (Spearman 1964).

RHINOCEROS HORNS

These differ from paired horns in being composed entirely of keratinised cells. Unlike other types of horns, the rhinoceros horn contains interwoven fibres similar in diameter to hairs, but quite different in structure. Between these fibres are horny cells which form packing material. These two morphological components are formed over different sites in the horn epidermis. This fibrous arrangement gives considerable mechanical strength. Very occasionally a horn is knocked off from the epidermis, but a new horn then grows again in the course of a year or so. Normal growth is slow and shedding does not normally occur.

ANTLERS

These are found only in deer and differ from paired horns in that entire structures are shed annually and regrow in the following spring. The epidermis over the bony core forms only a thin horny layer and numerous short hairs (velvet). De-novo development of new hair follicles occurs each spring, unusual in adult skin (Billingham, Mangold and Silvers 1959).

DWARF HORNS

The dwarf horns of the giraffe and okapi appear intermediate between paired horns and antlers, since the terminal caps are shed annually (Spearman 1964).

CLAWS, NAILS AND HOOVES

These tough keratinised appendages which develop over the dorsum of the digits in terrestrial mammals all have the same basic structure (Fig. 36), but vary widely in shape due to differential growth (Biedermann 1926). The dorsal portion of a claw, nail or hoof grows from the root which lies in an epidermal invagination. Claws are hardened by crystals of hydroxyapatite deposited in the cells in addition to containing cystine-rich keratin. Primate nails are normally not so strongly calcified (Pautard 1963; Jarrett and Spearman 1966).

The microscopical arrangement of the horny cells in a horse's hoof, and probably also in other hooves, differs from claws and primate nails in having an arrangement of fibres similar to that in rhinoceros horn. This gives great strength for galloping over hard ground.

Keratohyalin granules do not appear in the root, but a granular layer occurs beneath the thin horny layer (hyponychium) under the free edge of the primate nail. The latter forms the tough ventral side of a claw and the walking surface, sole plate, of a horse's hoof. Growth of these appendages is continuous and they are slowly worn away.

PANGOLIN SCALES

The dorsal skin in the scaly anteater or pangolin is covered with peculiar horny scales which are quite different in structure and composition from mammalian tail scales. Surprisingly, they resemble nails in their histological appearance, although the significance of this is obscure. In the armadillo, the dorsal plates are less strongly keratinised.

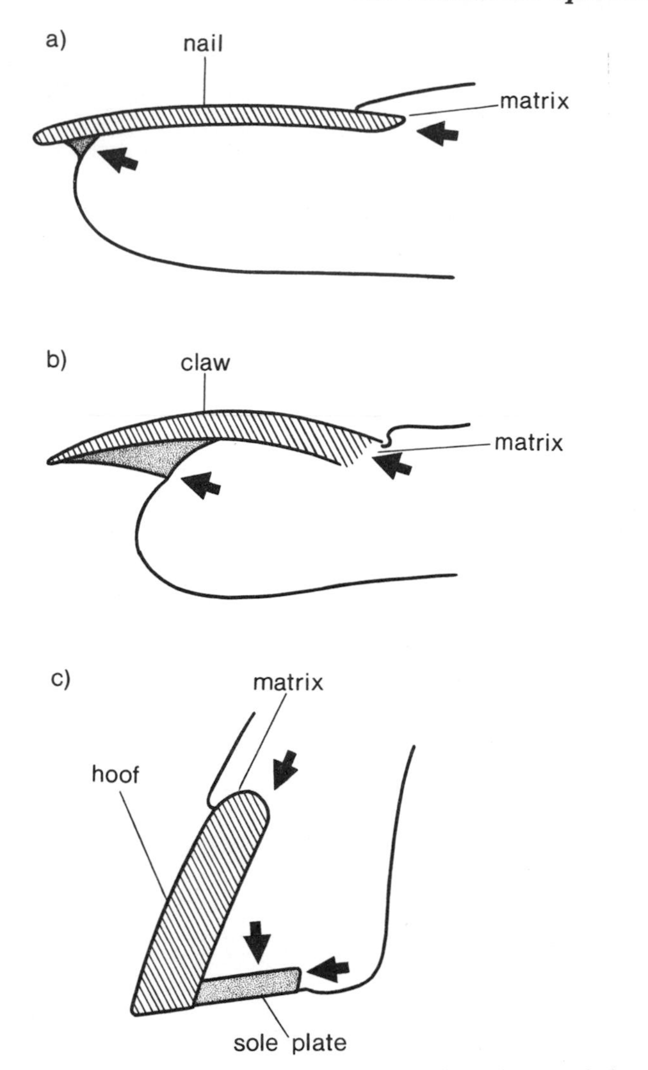

Fig. 36. (*a*) Primate nail; (*b*) mammalian claw; (*c*) horse's hoof.

BALEEN

This is the 'whalebone' which develops in long plates from the oral epithelium of the Mystacoceti and is used to sieve the small crustaceans on which the whales feed. Baleen is not bone, but contains a highly calcified keratin rich in hydroxyapatite (Pautard 1963).

SKIN GLANDS

SEBACEOUS GLANDS

These glands develop as epidermal outgrowths from the necks of hair follicles, and the secretory product, sebum, passes out through the hair canal to the skin surface (Fig. 32*a*). The growing hair, as well as the surrounding stratum corneum, is in consequence coated with a thin layer of this fatty material. Isolated sebaceous glands occur occasionally, such as around the nipples and lips, in regions where hairs were probably lost.

Sebaceous glands are holocrine, as in the avian preen gland, but they are much smaller appendages. Around the periphery of each gland are germinal cells which proliferate and cause an inward movement of more mature cells which later undergo autolysis and accumulate lipid droplets. Eventually the cells disintegrate and form sebum.

Composition of sebum

Human sebum contains cholesterol, its esters, squalene, waxes, triglycerides, free fatty acids, phospholipids, and a little protein. It can be dissolved in fat solvents, but is also a slow emulsifier and is therefore miscible with water derived from sweat.

Lanolin (wool fat), similarly miscible with water, is obtained from sheep sebum. Sebaceous glands occur in seals, where sebum may contain higher proportions of hydrophobic substances than in land mammals. Very little is known about species differences in sebum composition.

Functions of sebum

In terrestrial mammals it helps to keep the horny layer moist and supple under dry conditions, because of its water-absorbent properties. Dry keratin is brittle but, because of its affinity for water, is normally in equilibrium with the atmospheric humidity.

Control of secretion

Sebaceous glands secrete continuously and are unaffected by cyclical changes in hair growth. The glands are increased in size by androgenic hormones and shrink with oestrogens, which slowly affects secretion. Indirect influence comes through pituitary gonadotrophins (Montagna, Ellis and Silvers 1963). Although trophic nerves may occur, the glands are not under nervous control.

SWEAT GLANDS AND SCENT GLANDS (Fig. 37)

Two types of sweat glands occur in mammals which differ in their mode of development and method of secretion (Weiner and Hellmann 1960). The most constant in size and histological appearance are the eccrine glands,

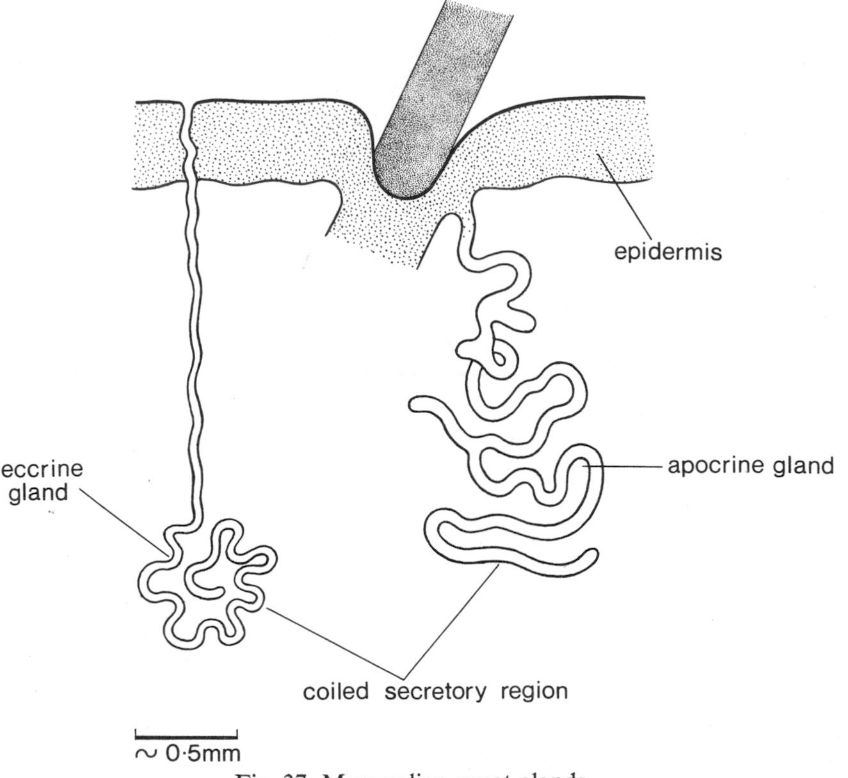

Fig. 37. Mammalian sweat glands.

derived as downgrowths from the epidermis. Eccrine glandular cells actively secrete sweat through microvilli into the duct lumen (merocrine secretion). The other type is the apocrine gland which is much more variable in size in different species. Thus, human apocrine glands are ten times larger than the eccrine glands, but in cattle they are smaller than human eccrine glands. Apocrine glands are derived in the foetus as outgrowths from the necks of hair follicles, although in adult skin, glands of this type sometimes occur in direct connection with the skin surface. In such instances, hair follicles may have atrophied or they get separated from the gland by differential growth after birth. Secretion in apocrine glands is characterised by budding off and loss of pieces of cytoplasm into the tubule lumen (apocrine secretion) or sometimes by complete cell disintegration (holocrine secretion), neither of which occurs in eccrine glands. Much more is known about eccrine gland function than about apocrine glands.

Eccrine glands

Each gland is made up of a long, unbranched, closely coiled secretory tubule (like a ball of string) which opens to the surface by a long excretory duct. Very occasionally, by developmental accident, two glands have a common terminal duct. In man, the coiled glandular portion is about 0.4 mm in diameter and lies deep in the dermis. In most mammals, eccrine glands are confined to the plantar and palmar skin where the watery secretion improves contact with the ground and is useful in primates for preventing slipping when grasping objects. Similar glands occur in the bare clasping region of the prehensile tail in New World monkeys. Elsewhere, eccrine glands occur occasionally in the moist nasal skin and are used in thermoregulation, but they are absent over the body generally except in higher primates and man, where they take on temperature regulation. Increase in numbers of eccrine glands is associated with a decrease in apocrine glands, although sometimes both types of glands occur, as in the human axilla and the trunk of the gorilla.

The glandular cells actively secrete a watery saline solution which involves the breakdown of glycogen in the cells to lactic acid. Water and sodium are later reabsorbed actively by cells lining the excretory duct, as in kidney tubules. Therefore, the sweat reaching the skin surface is more concentrated and the ionic balance is altered. Eccrine sweat is a slightly hypotonic solution of sodium chloride with some potassium, urea and lactic acid, but with no significant protein or mucopolysaccharides (Montagna 1962).

In man, the palmar and plantar eccrine glands are not used in temperature regulation and respond to emotional stimuli, which is also true to some extent of the glands in the forehead. Over the remainder of the body skin eccrine glands are closely involved in thermoregulation, and latent heat or evaporation from the skin surface lowers body temperature. Although eccrine glands secrete in response to a local increase in skin temperature, coordinated activity is through cholinergic endings of the sympathetic nervous system through the thermoregulatory centre in the hypothalamus. Palmar and plantar glands and the thermoregulatory glands all have cholinergic nerves, but the nerve pathways to the brain must be different.

Apocrine glands

These have a larger lumen than eccrine glands, and generally the excretory tubule has storage saccules branching off it. There are two distinct types: thermoregulatory glands which produce a profuse watery secretion, probably involving merocrine as well as apocrine secretion, seen in hooved animals, and accessory sexual scent glands which produce a scanty, viscous

secretion containing volatile substances and mucopolysaccharides. The secretion of apocrine glands is expelled at intervals in spurts by contraction of smooth muscle lining the storage sacs, which contrasts with the steady loss of fluid from eccrine glands.

The function of scent glands is to enable males and females to locate one another during the breeding season, particularly important in normally solitary animals which only pair up at times of mating and might otherwise never meet. They are most developed in species with a strong olfactory sense, when the volatile scent can be detected by other members over long distances. Scent glands are not confined to the urogenital skin. Thus, the musth glands behind the eyes in elephants are active during the breeding season in both sexes and are the only glands that have been found in the skin. Glands in the belly skin of the male musk deer secrete musk, used in the manufacture of all high-quality perfume.

Apocrine glands are poorly developed in human skin, but become active in the axillae and urogenital areas at puberty. Like the scent glands of animals, their functioning is dependent on sex hormones. In the skunk, these glands produce the pungent odour, a defensive mechanism. The mammary glands are also modified apocrine glands.

In the horse, thermoregulatory apocrine glands are hormonally controlled by adrenalin from the adrenal medulla in response to impulses received from the thermoregulatory centre. Although various different kinds of apocrine glands have been examined, none has been shown to be supplied by nerves. Rodents do not have either eccrine or apocrine thermoregulatory sweat glands, and many other mammals, including the dog and cat, have only very few sweat glands. Thermoregulatory apocrine glands occur mainly in the hooved animals (Artiodactyla and Perissodactyla) and in marsupials. The hippopotamus secretes a pink apocrine sweat, and sweat glands occur mostly in the fins of seals, and not at all in whales.

In view of the obvious wide variety of apocrine glands and differences in secreted material, it is unwise to generalise on mode of secretion or phylogenetic relationships. In particular, the thermoregulatory glands of hooved animals appear to be different from scent glands.

Poison glands

The duck-bill platypus is the only mammal with poison skin glands. The venom is secreted through an opening in a spine on each hind foot.

SUMMARY

Mammalain skin usually has a fairly thick dermis and subcutaneous layer, but bony plates rarely occur. In most species the dermis behaves in life as an elastic gel: combined properties of collagen, elastin and acid

mucopolysaccharides of the ground substance. More rarely, a much more fibrous dermis occurs.

The epidermis forms hairs which are new keratinised appendages, developed in follicular downgrowths from the epidermis. In phylogeny they probably arose as entirely new developments in the hinge regions between horny scales of the reptilian ancestors of mammals. Vertebrates show a general tendency to develop horny epidermal appendages, and the evidence is against the derivation of hairs from sensory appendages or reptiles or amphibians. The first hairs were probably tactile appendages, as in vibrissae, but they soon became involved in thermal insulation by development of secondary hairs to form the undercoat, necessary when mammals became warm-blooded. Another feature peculiar to mammals is the epidermal horny layer, the cells of which undergo much more complete autolysis than in other vertebrates and are left as little more than keratinised shells. This type of horny layer always has peculiar cytoplasmic granules of keratohyalin in the underlying transitional zone not found in other vertebrates. These granules also occur in mammalian footpad epidermis which has a specialised type of cornification. In both ontogeny and phylogeny keratohyalin granules and these mammalian types of horny layers occur in close relation to hair follicles and sweat gland development. The horny layers of lower animals are all of the basic parakeratotic-scale type, in which only limited autolysis occurs and nuclear remnants are often retained. This scale type of cornification also occurs in the tail scales of certain mammals and in modified form in aquatic species as well as in abnormal states.

Mammalian skin is supplied with numerous skin glands. Sebaceous glands develop from the hair follicle canals and secrete a lipoidal fluid which in terrestrial mammals is miscible with water. Its main usefulness seems to be to help keep the horny layer moist and pliable by admixture with absorbed sweat.

Sweat glands are of two types. The eccrine glands in the footpads of most mammals secrete a watery fluid which helps in gripping the ground in walking and for grasping objects in primates. In higher primates, especially man, these glands have spread all over the body and are used to lower body temperature by sweat evaporation from the skin surface. The thermostat is in the mid-brain and eccrine glands are supplied with motor nerves.

Apocrine glands vary much more widely, but are of two basic types: thermoregulatory glands in hooved animals and scent glands in various species, active during the breeding season. Neither type is innervated and secretion is effected by liberation of adrenalin from the adrenal medulla. The adrenal glands receive messages by nerves from the mid-brain thermostat. Scent glands produce a much more viscous fluid which contains volatile substances.

PART 2

COMPARATIVE FUNCTIONS

13

THERMAL REGULATION

A problem which all animals have to overcome if they are to survive is extremes of environmental (ambient) temperature (Allee *et al.* 1949; Causey-Whittow 1970). If this is below that produced in the animal by basal metabolism, heat is lost through the skin, and if the external temperature is higher than body temperature, heat is gained. However, because enzyme systems in cells work efficiently within only a narrow temperature range, any wide fluctuation in body temperature is disadvantageous, and often lethal in birds and mammals with their high metabolic requirements.

AQUATIC ANIMALS

These are buffered against extremes of ambient temperature by the surrounding water which, because of its high specific heat, in the tropics is cooler than dry land, and in polar regions is warmer than air. The low metabolic activity of many invertebrates together with their small size means that relatively little internal heat is generated. Also, their surface area for heat exchange relative to volume is large, so that small planktonic species have approximately the same temperature as the surrounding water. Marine invertebrates and fish nevertheless thrive and are able to breed only within narrow temperature ranges for each species, so that tropical and temperate seas have quite different faunas down to the first few hundred feet of water, but the deeper uniformly cold regions of the oceans have a more constant fauna. Warm-blooded vertebrates which inhabit polar seas, such as whales, seals and penguins, are insulated against heat loss by a thick layer of blubber. In water it is never necessary to cool the body by physiological means, for even in tropical seas the temperature never gets hot enough to cause a problem, but, if marine mammals are stranded on land in warm weather, heat stroke can occur.

TERRESTRIAL POIKILOTHERMIC ANIMALS

Cold-blooded (poikilothermic) animals are often thought to have the same body temperature as their surroundings. Certainly, body tempera-

ture in these animals varies more than in warm-blooded (homoiothermic) species, but many have evolved primitive mechanisms of thermal regulation and are not entirely at the mercy of the environment.

EXTREMES OF TEMPERATURE

Lowering of body temperature even to the freezing point of water over a short period is not always dangerous to an animal, except in mammals and birds. Cooling the tissues merely results in the slowing down of metabolism through tardier enzymatic reactions. Indeed, the bodies of invertebrates, fresh-water fish and even snakes can freeze solid in winter in temperate countries, and yet the animals survive to function normally again when they thaw out in the spring. Danger to life occurs when large ice crystals form in the cells and disrupt their contents, but small crystals cause no damage. A few mammals, such as the hamster, can withstand very low hibernating temperatures without ill effect. In bats, the body temperature drops to the ambient level when they are asleep during the daytime, but they are homoiothermic when active at night.

Much more dangerous than a temperature drop is a rise in body temperature, which may be sufficient to denature enzymes and kill the cells. The central nervous system is most sensitive to high temperature and neurones are the first cells to die.

THERMAL REGULATION IN INSECTS

Insects are the only fully terrestrial invertebrates and are able to regulate their body temperature quite efficiently by choosing optimum ambient temperatures and avoiding extremes. In cold weather, bees generate heat by beating their wings while at rest in the hive. This suggests a central nervous centre for heat regulation which determines the behavioural pattern. Insects have a well-developed temperature sense with thermoreceptors on the antennae and legs.

THERMAL REGULATION IN AMPHIBIANS AND REPTILES

In both amphibians and reptiles a thermosensitive centre is found in the hypothalamus, the precursor of the thermostatic control mechanism of warm-blooded animals (Bligh 1966). The few amphibians found in hot arid regions, and also reptiles, select optimum ambient conditions by hiding from strong sunlight and choosing the warmest patches in cooler weather. Some species can lower their body temperature by panting with evaporation of water from the mouth. Metabolic heat production, mainly by the

liver, can be varied to some extent and shivering occurs in some snakes exposed to cold, considerable heat being generated by these spontaneous muscular contractions.

TERRESTRIAL HOMOIOTHERMIC ANIMALS

Homoiothermic birds and mammals have a more efficient control of body temperature than poikilotherms. To maintain their high metabolic rates, small birds and mammals with a constant body temperature independent of the environment need to consume much more food than cold-blooded animals of the same size. Body heat is produced mainly through the oxidation of glycogen and fats in the liver and by muscular activity. Over one-third of the basal heat in mammals and birds comes from tension in uncontracted voluntary muscles. Shivering is only resorted to if additional heat is required. Some mammals, such as the Arctic fox and husky dog, are able to sleep without ill effect at ambient temperatures as low as −40 °C, because of their high basal heat production.

NEED FOR HEAT LOSS

In homoiotherms, mechanisms for both heat conservation and heat dissipation are required to control body temperature. In large tropical animals, such as the elephant, the skin surface area is small relative to the heat-producing volume, so that the main problem is sufficient heat loss in hot weather. This contrasts with small passerine birds which have a large surface area relative to volume, and the main problem is heat retention in cold weather. Adaptation to reduce heat loss is shown by those species subdivided into altitudinal clines, particularly common in birds: the smaller members of the species inhabiting warmer low ground and large members, with greater heat-producing volume relative to surface area, inhabiting colder high ground (Huxley 1963).

BODY TEMPERATURE IN BIRDS

The majority of passerine birds maintain a body temperature around 41 °C, higher than in most mammals. Since birds do not have sweat glands, loss of excess body heat is achieved by panting and evaporation of moisture from the mouth which cools the circulating blood in the oral mucosa and lungs. Other important regions of heat loss are the bare scaly legs of many birds. In very cold weather the blood vessels to the tarsal skin are collapsed which minimises heat loss, but in hot weather the legs become much more vascularised and more heat is lost. It is interesting that

the two sites where reptilian-like scales are retained in homoiotherms, the legs of birds and the tails of small mammals, are both sites of controlled heat loss. In very hot weather many birds spread and flap their wings to cool themselves.

Heat conservation is achieved very effectively by the plumage, feathers being better insulatory material than the hairy coats of mammals, important for small birds at cold ambient temperatures (Scholander *et al.* 1950). Still air is trapped beneath the down feathers and provides thermal insulation, but if either feathers or hairs become wet and the air spaces filled with water, the plumage or coat ceases to insulate the body. The closely arranged contour feathers, oiled from the preen gland, normally prevent the down from getting wet in aquatic species. Birds which spend much time in the water, such as ducks and geese, also lay down insulatory fat.

In cold weather the feathers are fluffed up by the reflex contraction of muscles attached to the follicles and this traps more air under the down. In contrast, in the ostrich and many tropical birds which have no down, erection of the contour feathers exposes the skin surface and heat is lost more readily.

BODY TEMPERATURE IN MAMMALS

Most eutherian mammals have a body temperature of 35–39 °C, but the more primitive marsupials maintain a body temperature around 30 °C.

Loss of heat occurs in terrestrial species in several ways. All, including man, use behavioural regulation and avoid extremes of ambient temperature. In small mammals with sweat glands confined to the pads, loss of heat is achieved by panting and by vasodilation of the more naked skin sites. This is true even of kangaroos and hooved animals which have thermoregulatory apocrine sweat glands operated by adrenalin, but in which sweat evaporation is used to lower body temperature only in extreme exercise. In these animals sweat is ejected in a single short spurt at times of extreme muscular activity, such as when escaping from some predator. These hormonally operated sweat glands, found in all large mammals except for the elephant, are therefore only a temporary means of dissipating excess body heat. Continuous neurally operated sweating, as in man, would in a species with a thick pelt merely make the fur wet, when surface evaporation from the coat surface would not lower skin surface temperature. In the presence of dry fur an apocrine spurt and evaporation from the epidermal surface lowers body temperature, but once the fur is wet, it ceases to do so. In naked human skin with its short hairs, continuous neurally operated sweating is more efficient and the major factor in lowering body temperature.

The opening up of blood plexuses in the dermis causes the circulating blood to be cooled in exposed sites such as the mouse tail, human skin, and the large ears of the elephant. Some species have vasodilator nerves, but in human skin the vasodilator polypeptide, bradykinin, is produced by active sweat glands. The ear pinna in many mammals is well vascularised, has only short hairs, and is an important organ of heat loss. Examples are rabbit and rat ears, the blood vessels of which increase in diameter when the skin is heated experimentally above normal body temperature.

SWEATING MECHANISMS

The adrenalin-operated apocrine spurt sweating mechanism and its control from the thermoregulatory centre in the hypothalamus through the adrenal medulla has been most closely studied in the horse (Weiner and Hellmann 1960).

Human eccrine sweating for temperature regulation is also controlled by this heat-regulatory centre (Montagna 1962). Thermoreceptor nerve endings in the dermis and voluntary muscles monitor body temperature, and blood temperature is also monitored as it passes through vessels near the brain centre. If the blood temperature rises above 37 °C, or the skin temperature (which is lower) above 34 °C, impulses pass to the nearby sweating centre also in the hypothalamus and messages are sent along nerves to the eccrine glands where acetylcholine is released and sweat is secreted. 0.58 kilocalories of body heat are lost per gram of evaporated water, the latent heat of evaporation which cools the body. Production of visible sweat which drops off the skin has no cooling effect. To work efficiently, the glands need to produce only sufficient water to evaporate as it reaches the skin surface. In the tropics, excessive sweat secretion is a sign of poor heat adaptation and is prone to occur in persons who fly out to a tropical country from a cold climate. Mineralocorticosteroids from the adrenal cortex promote salt and water resorption from the sweat excretory duct, as in the kidney tubule, which reduces the amount of fluid reaching the skin surface in adapted persons. In a non-adapted person, body temperature not controlled by profuse sweating rises rapidly and heat stroke may occur. The exhausted sweat glands then shut down and the skin becomes dry, but normal functioning returns if the person is placed in an air-conditioned room at a lower ambient temperature. Under extreme conditions 12 litres of water can be lost per day, and both the salt and water must be made good.

In a temperate climate continuous slow eccrine sweating does not commence in a nude person at rest until the atmospheric temperature rises to about 29 °C, but in muscular activity the heat generated raises the body temperature so that sweating occurs at quite low ambient temperatures.

In the humid conditions of a tropical rain forest, sweating and panting are of limited value, but here individuals are shielded from the radiant heat of the sun.

DESERT MAMMALS

Desert mammals (Maloiy 1972), such as the camel, Grant's gazelle and the red kangaroo, need to limit the use of panting and sweating so as to conserve water and salt. To achieve this, the thermostatic centre in the brain allows the body temperature to rise in the heat of the day to a higher level than at night. Thus, in Grant's gazelle, the body temperature reaches 46.5 °C at midday, but in Thomson's gazelle, a non-desert species, it does not go above 42.3 °C. Panting is resorted to in desert mammals only when body temperature reaches a high level. The brain temperature in Grant's gazelle is nevertheless maintained at a lower temperature than the rest of the body, because it appears that blood to the brain is cooled by inhaled air passing over a blood plexus at the back of the nasal cavity.

INSULATION BY MEANS OF FUR

Thermal insulation in terrestrial mammals is achieved by air trapped beneath the undercoat hairs (Barnett 1965; Spearman 1964). In whales, where fur is of no value, hairs have been lost. In cold weather, hairs are reflexly erected by arrector pili muscles, which increases the layer of trapped air as in plumage. The coats of different mammals vary widely in the density of their hairs and in the amount of underfur present, and therefore in their insulatory efficiency. In a small mammal, such as a mouse, because the animal would be impeded by the weight of the pelt, the hairs can never be long enough to provide efficient insulation under Arctic conditions, quite apart from the greater heat loss in a small animal because of the adverse surface/volume ratio. In consequence, small mammals, such as marmots, hibernate as winter approaches in very cold climates. It has been estimated that an animal has to be the size of an Arctic fox to carry a sufficiently heavy pelt to conserve enough heat when exposed to the Arctic winter. Surprisingly, the polar bear does not have a very good fur coat but is protected from extremes of cold by hibernation during the winter months (Scholander *et al.* 1950).

Many Arctic terrestrial mammals have a layer of subcutaneous fat to supplement the pelt, and human skin also has a layer of fat which helps to compensate for the absence of fur.

SUMMARY

Small animals with a large surface/volume ratio undergo considerable heat exchange through the skin. Insects and cold-blooded vertebrates have thermoreceptors and regulate body temperature by behavioural means. Warm-blooded birds and mammals utilise feathers and hairs for thermal insulation. In mammals sweat evaporation is sometimes used to lower body temperature. The ears and tail in mammals and the tarsus in birds are sites of controlled heat loss.

14

CHEMICAL AND NEURAL CONTROL MECHANISMS

CONTROL OF CELL DIVISION

The cycle of cell division is shown in Fig. 38. Replication of DNA occurs in mid-interphase. Uptake of H^3-labelled thymidine occurs during DNA replication and in consequence shows cells ready to divide. On the other hand, colchicine, used in counts of dividing cells, arrests cells in metaphase. These two methods therefore show different phases of the cycle (De Robertis, Nowinski and Saez 1965).

NATURAL INHIBITORS OF CELL DIVISION

Recently, natural inhibitors of cell division have been demonstrated in tissues. These substances, which Bullough (1962) termed chalones, are tissue-specific, but not species-specific, exosecretions of cells whose normal function is to control cell division in neighbouring cells. Frankfurt (1971) showed that rat epidermal chalone was even capable of inhibiting cell division in mouse epidermis stimulated to divide by powerful agents. However, surprisingly, hair follicles, although epidermal, are insensitive to chalones. These substances act by preventing DNA replication, and cells which have already formed new DNA are not stopped from dividing. So far work on chalones has been mostly in mammals, but they presumably occur in other animals. If a square piece of mouse skin is excised, the epidermal cells around the edges of the wound undergo rapid division, and Bullough and Laurence (1964) suggest that the reason for this proliferation is the loss of chalones from the wounded site. Clearly much more needs to be known about these inhibitors. In mouse skin, chalones are only functional in the presence of small amounts of adrenalin. There may be an effect on the enzyme adenyl cyclase which converts ATP to cyclic AMP since it accumulates in interphase at the time of chalone action. Some even suggest that chalones may be forms of adenyl cyclase.

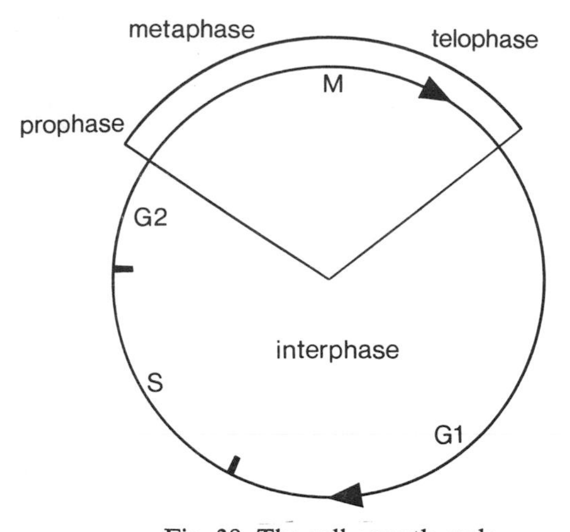

Fig. 38. The cell growth cycle.

HORMONAL CONTROL OF INVERTEBRATE ECDYSIS

MOULT IN ARTHROPODS

Insects

Ecdysis in Arthropods is hormonally controlled and larval moult has been investigated in most detail in the hemipteran bug *Rhodnius*. In many insect larvae, moult follows a large meal, which stimulates neurosecretory cells in the corpora cardiaca in the neural cerebral region, to secrete a hormone which prepares the prothoracic glands so that after a few days they begin to secrete thc moult hormone ecdysone. The neural hormone does not itself trigger off ecdysone secretion, but it is necessary for its production. In addition, juvenile hormone (neotenin) from the corpora allata, an epithelial endocrine gland situated behind the brain, is required for ecdysone secretion. If the allata are removed, the next moult takes place, but the prothoracic glands atrophy and no further cuticle shedding occurs. In the primitive wingless Thysanura, periodic ecdysis occurs throughout life, but other adult insects do not moult. This is because the allata are lost during metamorphosis and the prothoracic glands regress. If the haemocoel of a larva containing allata is grafted to that of an adult insect (parabiosis), neotenin from the larva reactivates the prothoracic glands of the adult. Ecdysone is then secreted and ecdysis occurs in the adult. Ecdysone has been chemically isolated and prepared in crystalline form (Wigglesworth 1964).

Crustacea

The Y organ, a non-neural gland situated in each antennary segment, secretes a moult hormone chemically similar to ecdysone, and appears homologous with the insect prothoracic glands. Crabs in which Y organs have been removed cease to moult. Nevertheless, if the moult cycle is already past its early stage when the Y organ is removed, ecdysis is completed normally. This shows that the hormone does not affect the later stage concerned with calcium resorption. Ecdysis is restored by implanted Y organs. Moulting in adult crabs and lobsters continues into old age, but eventually ceases. The activity of the Y organ is controlled by a hormone secreted by the neural-sinus gland, a neural endocrine gland situated in each eye stalk and comparable to the insect cardiaca. After terminal moult, the Y organs atrophy like the insect prothoracic glands (Carlisle and Knowles 1959).

Influence of light on moult

Intensity and duration of daylight acting naturally through the eyes, probably on the cardiaca or sinus glands, influence the hormonal control of moult in arthropods, but requires further elucidation.

Moult in other invertebrates

Control of cuticle ecdysis in other arthropods, in nematodes (Bird 1971), and in annelid worms, as well as in other invertebrates, is most probably also hormonally controlled, but not enough is known of the endocrinology of these animals to comment further at present (Barrington 1963).

HORMONAL CONTROL OF FISH SKIN

Thyroid hormone stimulates the deposit of guanine in guanophores of the salmon and speeds the rate of scale formation in sturgeons, but little is known of the effects of hormones on epidermal mucus secretion or on cuticle ecdysis in fish, although probably the latter is hormonally controlled.

Breeding tubercles

Gonadectomy in the goldfish prevents development of these keratinised tubercles and causes those already present to regress. Injected oestrogens cause tubercles to form in males of the blunt-nosed minnow, while testosterone causes tubercle development in both sexes. Distal pituitary extracts also stimulate tubercle formation.

In male goldfish, tubercles appear within two weeks of increasing duration of daylight from eight to twelve hours, and the tubercles are moulted three to four weeks after a decrease in daylight from twelve to eight hours.

Males maintained at twelve hours of light daily developed tubercles after four months, but those kept in sixteen hours of light failed to develop tubercles at all (Wiley and Collette 1970).

The likely mechanism for tubercle development is that light falling on the retina causes nerve impulses to pass to the hypothalamus in the brain. The distal pituitary lobe is then stimulated to secrete gonadotrophins which cause the gonads to produce oestrogen and testosterone, both of which stimulate tubercle development. Increasing duration of daylight during the spring, and decreasing duration in autumn, is the natural stimulus to tubercle development in temperate regions.

The horny caps of lamprey teeth are replaced regularly, but the possible hormonal control mechanism has not been determined, and replacement is not affected by hypophysectomy (Larsen 1973).

MOULT IN AMPHIBIA AND REPTILES

AMPHIBIAN EPIDERMAL CELLS

Sloughing of the horny layer occurs once every few days to once every few weeks in different species. As in other cold-blooded animals, metabolic activity is affected by changes in ambient temperature. In very cold weather moulting ceases, and there is more rapid replacement in hot weather. Light is probably important for pituitary stimulation. Periodic shedding of the superficial epidermal cells is seen more readily after metamorphosis when keratinisation occurs, but loss of epidermal cells also occurs in unkeratinised skin associated with cuticle ecdysis.

There is an inherent rhythm in the epidermis of cell proliferation and desquamation or moult. Hypophysectomy in toads prevents moulting, although keratinisation continues so that a thick horny layer is formed.

The distal pituitary lobe elicits moulting in the tailed urodeles through the effect of its thyrotrophic hormone on the thyroid. Extirpation of either of these endocrine glands results in termination of moult which can be restored by hormone therapy.

In frogs and toads, there is a different hormonal control mechanism, in which the distal pituitary lobe stimulates the adrenal cortex to secrete mineralocorticoids, which elicit moult.

Thyroxine triggers off metamorphosis, and in consequence epidermal keratinisation in both Anura and Urodela, but it does not elicit moult in frogs and toads. Neither do mineralocorticoids elicit moult in newts. The sex hormones have not been shown to influence moulting in Amphibia. One action of hormones on moult in Amphibia may be on the junctions between the horny cells. Thus, in hypophysectomised amphibians, the dorso-ventral junctions between the horny cell layers are not broken down as normally occurs (Larsen 1973).

SLOUGHING IN LIZARDS AND SNAKES

Sloughing in the lizard *Gekko gekko* is also initiated by the thyroid controlled by the distal pituitary thyrotrophic hormone. The control of cyclical changes in keratinisation is little understood. This involves the regular alteration in keratin synthesis from beta to alpha keratin, and a decrease in cystine content until in the nucleated layer (fission plane) no keratin is laid down at all. Probably there is an intrinsic rhythm involving cell proliferation and keratinisation which is hormonally regulated, as in Amphibia, but the exact points of hormonal action and effects of light in Squamata remain to be determined (Bellairs 1969).

MOULT IN BIRDS

CONTROL OF FEATHER GROWTH

Plumage moult in birds has been most extensively studied in the domestic fowl and in pigeons (Voitkevich 1966). The renewal of feather growth in spring, which precedes prenuptial moult, is in temperate regions initiated by the increasing daily duration of light falling on the retina during still short periods of daylight. The autumn moult occurs during shortening light periods and, as in fish, actual duration of daylight is much less important. Nerve impulses to the brain hypothalamus determine the rate of secretion of first thyrotrophic hormone and later gonadotrophins by the distal pituitary lobe. A high level of thyroxine induces growth (anagen) in most resting follicles and so indirectly causes moult because the old feathers are pushed out mechanically by formation of new feathers. However, the chick down feathers and adult flight feathers are independent of thyroxine. Removal of the thyroid in adult birds therefore causes atrophy of all feather follicles except for flight feathers, the endocrine control of which is uncertain. Oestrogen promotes ovulation but retards feather formation, and moult and egg-laying therefore occur at different times. Testosterone has a weak inhibitory effect on feather growth. In both sexes, gonadectomy leads to continuous renewal of feathers and frequent moulting without the resting phases of normal birds.

In passerine birds, increasing or decreasing duration of daylight combined with changes in light intensity stimulate feather growth and consequent moult of the old plumage. Later, oestrogen levels increase and the follicles enter the resting (telogen) phase, at which time the eggs are laid.

CONTROL OF FEATHER FORM

The form of the feathers is also influenced by hormones (Sturkie 1965; Spearman 1971). The lower part of the feather follicle in the domestic fowl

is thyroxine-dependent, but the distal region is strongly influenced by oestrogen, which in most birds determines female plumage characters. Male plumage occurs in the absence of oestrogen and is not affected by testosterone.

Male hormone in most species has little effect on plumage morphology in normal physiological doses. However, in the Campine and Sebright breeds of the domestic fowl, female plumage is induced by both oestrogen and testosterone, which explains the hen-type feathering in cocks of these breeds.

INTRINSIC RHYTHMS

Each follicle has an intrinsic rhythm of growth (anagen) and rest (telogen), and onset of growth within a pteryla is determined by a local hormone produced by follicles in the growth centre. Hormonal control of plumage renewal is largely by inhibiting the rate of growth, and long resting periods occur when feather formation is suppressed.

BALANCED GROWTH

The bilateral balanced growth of feathers on the two sides of the body, important for flight, is unlikely to be under endocrine control, and it has been suggested that trophic nervous control of follicle growth may be important.

MAMMALIAN HAIR GROWTH

CONTROL OF HAIR GROWTH

Mammalian hair follicles also show an intrinsic rhythm of growth (anagen) and rest (telogen) and have a catagen stage of atrophy not seen in birds. Anagen involves renewed growth of the follicle germinal cells at the base of the permanent part of the follicle. When eventually the old hair is moulted, this is due to loosening of its fibrillar attachment to the follicle. Club hairs are lost more rapidly in female rats in which the ovaries have been removed, and implanted oestrogen reduced hair loss. Certainly, therefore, hormones do influence loss of club hairs, although how this happens is not clear. Since the new hair grows up alongside the old hair, moult is not a simple mechanical process as in plumage replacement.

Many wild mammals in temperate regions moult twice a year, in the spring after the young have been weaned, and again in the autumn. Examples are the lemming, hare and mink. A twice-yearly replacement is necessary to provide a long winter pelt and short summer coat, sometimes differently pigmented, for seasonal requirements. In the Bactrian camel, the winter coat is particularly long and the summer coat very short, so

that animals appear much larger in winter. Some other mammals, such as the coyote and common seal, have only the spring moult, and others, such as the southern elephant seal and the northern fur seal, have only autumn moults (Ling 1970).

Moult in most temperate species is controlled by lengthening periods of daylight in the spring and shortening periods in the autumn linked to the reproductive cycle, as in other vertebrates, and has been studied in most detail in the mink (Bissonnette and Wilson 1939). The photoperiodic pathway is through the eyes. Nerve impulses to the hypothalamus stimulate the distal (anterior) pituitary lobe, which controls the hormonal activity of the thyroid and gonads (Donovan and Harris 1954). Nocturnal species, or those which live most of their lives under cover from daylight, such as the house mouse *Mus musculus* and common rat *Rattus norvegicus*, have lost this response to light periodicity, and frequent moults occur throughout the year; this is also true of laboratory strains. Neither is human hair replacement or hair growth in tropical mammals obviously influenced by light.

Although all species have a genetically determined intrinsic rhythm of hair growth and rest, the cycle is delayed or speeded up by hormones. Most experimental studies on hair growth have been done on laboratory mice, rats and rabbits with prominent single-wave hair cycles. Length of the growth period has not been shown to be hormonally controlled, but the rate of hair growth is reduced by oestrogen in rats so that spaying increases hair length. Probably for this reason hairs in male rats are slightly longer than in females. Initiation of anagen in telogen hair follicles is induced in rats by thyroxine as it is in feather follicles, while oestrogen and testosterone and glucocorticoids delay anagen. Hypophysectomy reduces the duration of telogen.

CONTROL OF HAIR FORM

Different types of hair follicles in the same individual sometimes respond to hormones in different ways. Thus, in man at puberty, testosterone promotes growth of longer and thicker beard hairs in males, but the scalp hairs are not affected or are even reduced in size (in genetically predisposed individuals with male pattern baldness). A change in hair form from the infantile coat to that of the adult occurs in some mammals when sexual maturity is reached, and is due to similar sex-hormonal changes. Examples are the loss of the curly coat in young seal pups, replaced by straight hairs, and loss of bristles in Merino lambs on reaching maturity.

The endocrine control mechanism of the hair cycle is therefore through the distal pituitary lobe by means of gonadotrophic, adrenocorticotrophic, and thyrotrophic hormones, with consequent secretion of oestrogen,

testosterone, glucocorticoids and thyroxine. Cyclical changes in epidermal thickness are similarly determined (Ebling and Johnson 1961; Ebling 1965).

OTHER MAMMALIAN KERATINISED APPENDAGES

The growth and shedding of horns in the pronghorn antelope and of antlers in deer are probably hormonally dependent. Paired horns and antlers are usually largest in males and are often absent in females. In sheep an autosomal gene for horn-bearing is dominant in males and recessive in females, the expression being determined by the sex-hormone balance (the genes show sex limitations).

VERTEBRATE MOULT CYCLES

Vertebrate epidermal moult cycles from fish to mammals are influenced by changes in duration and intensity of daylight falling on the retina, and this stimulates the distal pituitary lobe to secrete trophic hormones which stimulate other endocrine glands, the thyroid, adrenal cortex and gonads, responsible for the final stage of the control mechanism. The ultimate hormonal control is not at all constant, and hormones responsible for growth and moult differ from group to group, while sometimes opposite effects occur in different animals. The photoperiodic neural pathway through the brain is primitive, and occurs also in control of arthropod ecdysis and of skin colouration in both invertebrates and vertebrates.

GLANDULAR ACTIVITY

Only mammalian sebaceous glands have been extensively studied. Their intrinsic secretory activity is decreased by oestrogen and increased by testosterone and adrenocortical androgens; the latter are found also in females (Montagna, Ellis and Silvers 1963).

Resorption of water into the body from the eccrine sweat-gland tubules in man and inward movement through the epidermis in Amphibia depend similarly on active transport of sodium and chloride. Both are promoted by mineralocorticoids such as aldosterone. Water resorption is increased by the pituitary neurohypophysial antidiuretic hormone (Montagna 1962; Davson 1970).

THE CONTROL OF PIGMENT CELLS

PHYSIOLOGICAL COLOUR CHANGE

With the exception of cephalopod chromatophores controlled by nerves acting on extrinsic muscles, all rapid physiological colour changes, whatever the type of pigment, are affected in the same manner by altering the

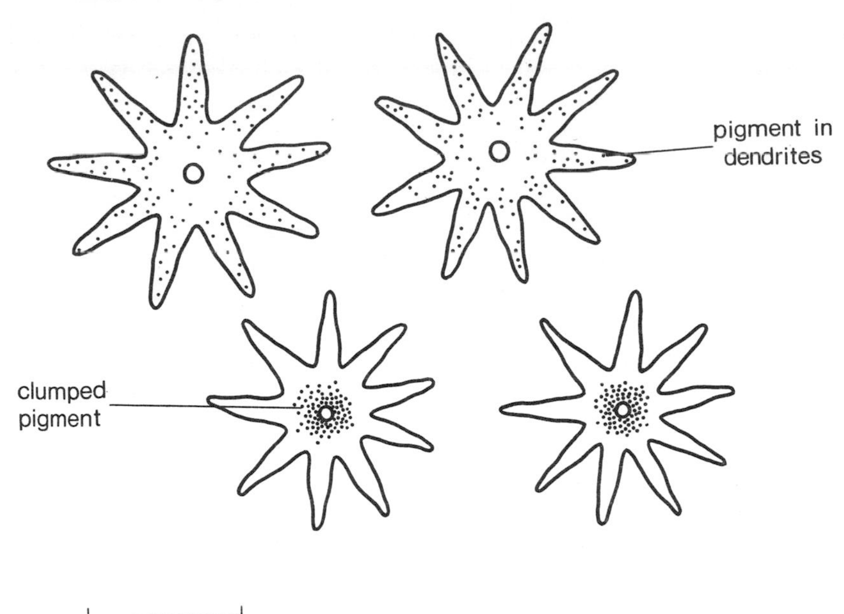

Fig. 39. Surface view of fish chromatophores with clumped and dispersed pigment granules.

distribution of the pigment granules within the cytoplasm by changes produced within the cell itself. If pigment is clumped around the nucleus, wide unpigmented areas of skin appear between the chromatophores and the integument is light coloured. If pigment extends into the dendritic processes, which often interdigitate with neighbouring pigment cells, the skin appears dark (Fig. 39).

CONTROL OF CHROMATOPHORES

The control of pigment movement in the chromatophores is by direct light stimulation in echinoderms and some other invertebrates. In other groups, including crustaceans and vertebrates, light acts through the eyes and brain and stimulates the secretion of hormones, or direct nervous stimulation occurs which affects the chromatophores. Thus, in decapod crustaceans, the sinus glands release a hormone which affects the pigment cells. Depending on the pigment, it either concentrates it and lightens the skin, or disperses it, making the skin darker. Other hormones produced by various neurosecretory cells are also important in evoking responses in chromatophores. Crustacean chromatophores are considerably more complex than those of vertebrates and often different; brown, red, yellow

and blue pigments occur in the same cell with dispersion of these pigments, each into a different dendrite. Sometimes the position is further complicated by the presence, as in the shrimp *Crangon*, of no less than eight different types of chromatophores (Waring 1963; Fingerman 1969).

EVOLUTION OF CHROMATOPHORES

The most primitive chromatophore stimulatory system was probably direct light stimulation, then came endocrine stimulation and direct nervous control. In crustaceans, in some fishes, in amphibians, and in many reptiles, pigment dispersion is controlled by hormones, but in widely scattered groups, notably cephalopods, leeches, some teleost fishes, and in the true chameleons among reptiles, nervous control is dominant, possibly even exclusive. Melanophores, up to and including those in reptiles, continue to respond to direct light stimulation in addition to having central control.

CONTROL OF VERTEBRATE CHROMATOPHORES

Most teleost fish examined have melanin-clumping sympathetic nerves with several neurones to each melanophore, and melanin-dispersing nerves are controversial. Among elasmobranchs, only pigment-clumping nerves occur in the dog-fish, *Mustelus* and *Squalus*, but in the skate, *Raia*, in another genus of dogfish, *Scyllium*, and in the lamprey, *Lampetra*, there is only hormonal control. Skin darkening is caused by the intermediate pituitary lobe melanocyte-stimulating hormone (MSH) which disperses melanin in the dendrites; also involved in melanophore control in Amphibia and in many reptiles which show colour change. Cholinergic nerves clump melanin in Amphibia. In addition, a melanin-clumping hormone occurs in *Raia* and *Scyllium*, possibly melatonin, which in frogs is secreted by the pineal gland and lightens the skin. One wonders whether the pineal third eye in *Sphenodon* may be concerned in stimulation of melatonin formation in response to light changes!

In true chameleons, melanophores have sympathetic nerves which clump melanosomes, and dispersing nerves are controversial. In some other lizards clumping is caused by circulatory adrenalin, and possibly by thyroxine or melatonin.

In the frog, both dermal and epidermal melanophores are darkened by MSH, but only the dermal melanophores are lightened by melanin-clumping agents. Amphibian and reptilian epidermal melanocytes are similar to those of mammals and birds in that the epidermal basal cells phagocytose the tips of dendrites containing melanosomes, so that pigment is transferred to the epidermis. Physiological colour change does not

occur in either birds or mammals which are unaffected in this way by hormones or other agents. Nevertheless, the increased epidermal pigmentation produced in mammalian epidermis after injections of MSH may be due in part to dispersion of pre-melanosomes into the melanocyte dendrites where they can be phagocytosed more readily by epidermal cells.

Hormonal pigment change is relatively slow to take effect after light stimulation of the retina, and it often takes several hours for a colour change to occur. In contrast, neurally affected colour change is much more rapid and often occurs within a few minutes, as in the plaice *Pleuronectes* (Montagna and Hu 1967; Waring 1963).

CONTROL OF MELANOGENESIS

The hormone MSH from the pars intermedia of the mammalian pituitary also stimulates melanin formation. Local differences in pigmentation are also affected by sex hormones, and the adrenal cortex exercises some effect on melanogenesis (Montagna and Hu 1967; Chavin 1969; Riley 1972). In arthropods, ecdysone activates tyrosinase and Dopa decarboxylase involved in melanogenesis and sclerotisation.

THE INDIRECT ACTION OF SOME HORMONES ON CELLS

Recent work has shown that many hormones, for example adrenalin, act indirectly on cells by activating the enzyme adenyl cyclase which is bound to the plasma membrane and converts ATP to cyclic AMP. The latter, by transfer of phosphate ions, determines the chain of enzymatic reactions in the cell associated with the hormonal effect. As mentioned earlier, chalones possibly act in this way.

SUMMARY

Light periodicity through the mediation of the brain and endocrine glands is important in controlling both cuticle ecdysis in arthropods and moult of keratinised epidermal cells in vertebrates. Probably the latter process was superimposed on the more primitive cuticle ecdysis. Physiological colour change is effected by a similar pathway through the eyes. Slower colour changes are produced by hormones and more rapid changes by nerves to chromatophores. The vertebrate pituitary gland plays a central role in these endocrine mechanisms.

15

COMPARATIVE SYNTHETIC PROCESSES

THE RANGE OF PRODUCTS

The products of epidermal cells include both exosecreted and retained substances in a variety of fibrous and globular proteins, chitin, mucopolysaccharides, lipids and waxes. Toxic substances and aromatic scents are produced by epidermal glandular cells.

The dermal cells exosecrete fibrous proteins, glycoproteins, mucopolysaccharides and cellulose. Oxidation products of tyrosine include quinone polymers and melanin. Mast cells form histamine and other amines.

In addition, skin cells secrete in much smaller quantities enzymes and chalones which are protein in nature.

As an integral part of the cell, there are contractile actin filaments and skeletal fibrils, while lipoproteins form membranous structures.

THE GENETIC CODE AND PROTEIN SYNTHESIS

In an animal cell the nucleus is separated from the cytoplasm by a triple-layered membrane some 200 Å thick. During cell division this breaks down into fragments, reformed into a membrane at the end of mitosis. The nuclear spindle fibres which guide the movement of chromosomes in anaphase are formed and broken down again during mitosis. Lysosomal enzymes, including acid phosphatase, released in dividing basal cells, enter the nucleus and are probably involved in these breakdown processes.

The nucleus contains most of the deoxyribonucleic acid (DNA), the genetic material which controls protein synthesis. Each chromosome contains along its length a pair of twisted polynucleotide chains: the double helix of DNA which has a molecular structure like a spiral ladder with the uprights the phosphate and sugar parts of each nucleotide, and the rungs the purine and pyrimidine bases on adjacent chains cross-linked by hydrogen bonds. The bases in DNA are adenine, cytosine, guanine and thymine (A, C, G and T). Pairing is specific, so that A on one chain fits only T on the adjacent chain, and G with C. The base sequence along each polynucleotide constitutes the genetic code for protein synthesis and acts as a template for transcription of ribonucleic acid (RNA), which in turn codes the sequence of amino acids in each protein.

DNA REPLICATION

When new DNA is formed in interphase, adjacent chains in each chromosome unwind and separate by breakdown of hydrogen bonds. Nucleoside triphosphate units free in the nuclear sap then link with bases on the separated DNA chains and are joined together to form two new lengths of DNA, catalysed by the enzyme DNA polymerase. Just prior to mitosis, the cell has therefore four instead of two DNA strands to each chromosome, later reduced to two strands by chromosomal division. In this way the genetic code in the fertilised egg is distributed to all cells of the body.

The code is read from one end only of the DNA chain to the other in triplets of bases without overlapping. Since a combination of three bases in sequence codes for one amino acid, there are theoretically sixty-four kinds which can be coded, although only some twenty occur in nature. A DNA base sequence of TTT, CCA, and AGA is the code for the tripeptide with the amino acid sequence lysine, glycine, serine.

RNA TRANSCRIPTION

Transcription of RNA on DNA is similar to DNA replication in that it is elaborated according to the DNA base sequence. Four different kinds of bases also occur in RNA nucleotides, but one, uracil, is present only in RNA (A, C, G and U) and another, thymine, is confined to DNA. When RNA is transcribed, U combines with A in DNA; A combines with T in DNA and G with C or C with G in DNA. Joining of RNA nucleotides into a chain is catalysed by the enzyme RNA polymerase. Transcription of RNA was originally worked out in bacteria which do not have a nuclear membrane. Here the code is transcribed from the DNA to make a specific messenger (m) RNA. The bacterial messenger RNA is then detached and passes to the ribosomes where it acts as a template for the end-to-end arrangement of amino acids in the polypeptide chain to be synthesised. In animals, nuclear RNA does not enter the cytoplasm and so the mRNA which reaches the ribosomes is presumably retranscribed at the nuclear membrane from nuclear RNA formed on DNA. Cytoplasmic mRNA constitutes about 5 per cent of RNA and the life of a particular molecule is extremely transient. Most cytoplasmic RNA is a structural constituent of ribosomes and possibly acts as a guideline for the orientation of mRNA. Appreciable intranuclear RNA is found in the nucleolus during protein synthesis.

POLYPEPTIDE SYNTHESIS

Free amino acids which enter into the cell are shepherded to their correct place on the mRNA chain in the ribosome by units of transfer (t) RNA;

short chains which each link by covalent bonds to one particular amino acid only. In the ribosome, the different tRNA units link with mRNA in the coded sequence which in turn determines the amino acid sequence. The covalent bond between the first unit of tRNA and its amino acid breaks and the carboxyl group of the latter then forms a peptide bond with the amino group of its neighbour, catalysed by the enzyme, peptide synthetase, with energy provided by the breakdown of ATP. Further amino acids are added in sequence, so that a polypeptide is formed (the primary protein structure). Long protein chains require correspondingly long mRNA molecules. If one ribosome is not long enough, several are strung together by a strand of mRNA to give a polyribosome. Ribosomes are permanent organelles and at different times can be used to code for different proteins. Probably more than one protein may be coded on different mRNA strands attached to the same ribosome.

SECONDARY AND TERTIARY MOLECULAR STRUCTURE

The secondary structure of a protein defines the way a polypeptide chain is arranged in an alpha helix, in the beta-pleated form, or closely coiled as a globular protein. This is determined mechanically by possible bonds between different points on the chain and on sites of repulsion. Important linkages in this respect are hydrogen bonds, polar linkages and Van de Waals forces.

Polypeptide chains, each with a different amino acid composition, are sub-components of much larger protein molecules such as keratin or haemoglobin. The various polypeptides link together after leaving the ribosomes by a simple mechanical process referred to as complementation, involving the forces just mentioned. This does not require energy, and once the subunits are released in the right proportions they rapidly find their partners with reactive groups in the whirl of molecular collisions which occur in the cell. The way these different subunits are arranged, as in the supercoiling round one another of keratin microfilaments, themselves made up of polypeptide helical filaments, is the tertiary structure of protein. It is unlikely that complementation of different subunits (quaternary structure) could occur outside the confines of the cell. Aggregation and bonding together of like subunits can also occur, as in collagen, and readily takes place outside cells.

Those cells which retain synthesised protein, such as mammalian epidermal cells, have mostly free ribosomes, and the protein is deposited in the cytoplasm. Others, such as many invertebrate epidermal cells and fibroblasts, have most ribosomes attached to the endoplasmic reticulum which appears granular in electron micrographs. This is an extensive system of ultrastructural lipoproteinous membranous tubules which leads to the

Golgi apparatus, a collection of vesicles near the nucleus. Isotope studies show that proteins formed on the bound ribosomes pass along these tubules to the Golgi apparatus. From there, the protein passes either along other tubules to the plasma membrane where it is exosecreted, or it is budded off from the Golgi apparatus into small individual membranous vesicles, such as lysosomes. Hydrolytic enzymes are stored in these lipoproteinous bodies to be released at a later stage (Ambrose and Easty 1970; Shultze and Hcrcmans 1966).

TYPES OF EPIDERMAL CUTICULAR PROTEINS

Various exosecretions make up the cuticles of invertebrates and lower vertebrates. Two main types of cuticular protein are known to occur in different animals, but others may also exist. Fibroin proteins, which are sometimes quinone-tanned, are found in arthropod cuticles (Pryor 1962), and collagen, which is also sometimes quinone-tanned, occurs in nematode and annelid cuticles (Bird 1971). In most other animals the nature of the fibrous protein present is unknown, although probably one is always necessary to provide tensile strength. Whitear (1970) found that in fish cuticles the ultrastructure often varies from site to site, which is true also of other animals, such as in the joint and sclerite regions of arthropods. Fibrous material also passes out of the cells into the intercellular space where it contributes to the fish cuticle and to the material between the keratinised cells of mammals (Brody 1966). Similar cuticle-like material occurs in human foetal skin. In fibroins some 50 per cent of the amino acids present are glycine, alanine and serine. Several different fibroins are synthesised by the epidermal cells and when stabilised by quinone cross-linkages they become tough insoluble sclerotins. Arthropod cuticular fibroins occur as beta chains.

RESILIN

This untanned elastic fibroin protein occurs in the cuticle only in the joints of certain arthropods, and is chemically different from the sclerite fibroins. In the unextended state, resilin is non-birefringent as the molecule is randomly coiled, but when extended it becomes strongly birefringent, indicating molecular orientation. The chains are cross-linked by two rare aromatic amino acids: one a diaminodicarboxylic acid, and the other a triaminotricarboxylic acid (Andersen and Weis-Fogh 1964).

MICROVILLI

The cuticular surface of exosecretory cells is increased in area by microvilli, also seen in human foetal epidermis. Sometimes, as in annelids, pogonophores and fish, the crypts contain secreted fibrils.

CHEMISTRY OF QUINONE TANNING AND ITS RELATION TO MELANOGENESIS

SCLEROTINS

These occur in certain invertebrates but not vertebrates, and owe both their stability and brown colour to quinone cross-linking of widely spaced polypeptide chains. These quinone polymers originate separately from the cuticular fibroin chains between which they are later interpolated, built up from the amino acid tyrosine in a series of oxidative stages. Thus, there is an increase in blood tyrosine in the larva of the blow fly *Calliphora* just prior to ecdysis. Extracellular tyrosine is oxidised to 3,4-dihydroxyphenylalanine, then to dopaquinone which is decarboxylated and acetylated to *N*-acetyldopamine. The various oxidative stages are catalysed by tyrosinase and other phenolic oxidases which act extracellularly. Decarboxylation and acetylation are also affected by specific enzymes. Polymerisation then occurs to give a bulky orthoquinone polymer or catechols are formed which link either to amino side groups or sulphydryl groups of the protein chains (Figs. 40*c* and 41). The cuticular tanning process is related to melanogenesis. Indeed a pigment is formed by both processes, and the two diverge only after formation of dopaquinone. In the absence of free phenolic substances, tyrosine attached to the protein chains may possibly be oxidised to quinones (Brown 1950; L'Hélias 1970).

EXTRACELLULAR MELANIN

In melanin formation, condensation through the tyrosine amino groups gives an indole ring not found in sclerotins. In both vertebrates and invertebrates, indole melanin in melanosomes occurs as a cytoplasmic pigment, but it is also formed as an exosecretion in some insect cuticles. The enzyme tyrosinase and melanin formation can occur both inside and outside cells.

INTRACELLULAR MELANIN

This brown pigment is a complex, highly stable, and insoluble polymer of indole-5,6-quinone. It is formed in cytoplasmic organelles, the melanosomes. Completed melanin is closely linked to the protein strand of the

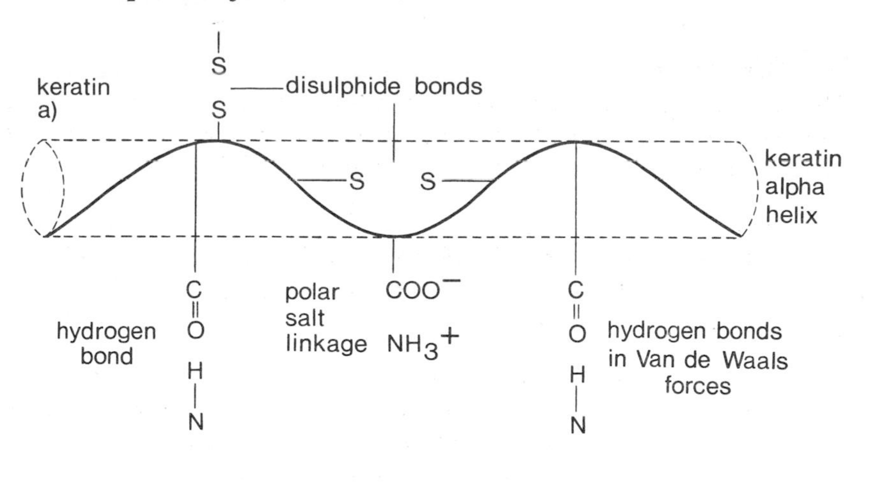

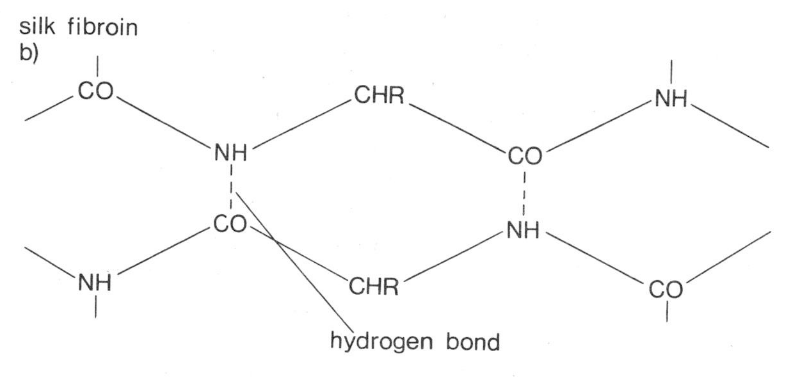

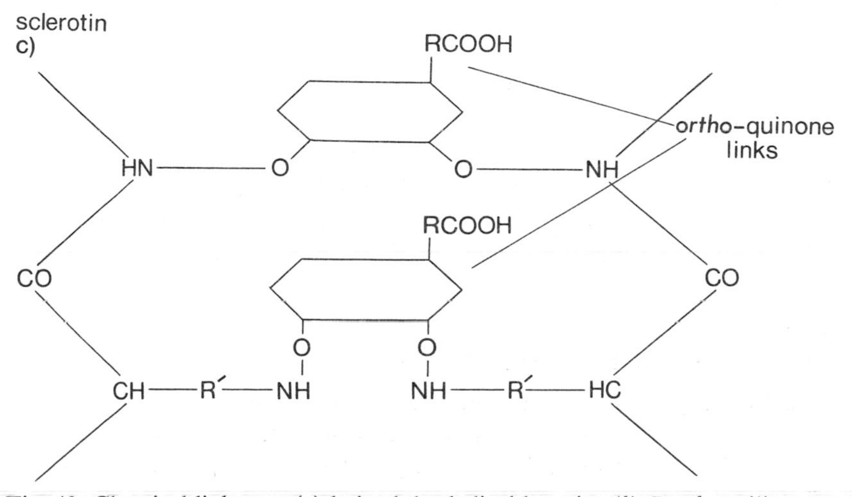

Fig. 40. Chemical linkages: (*a*) hair alpha-helical keratin; (*b*) *Bombyx* silk; (*c*) cuticular sclerotins.

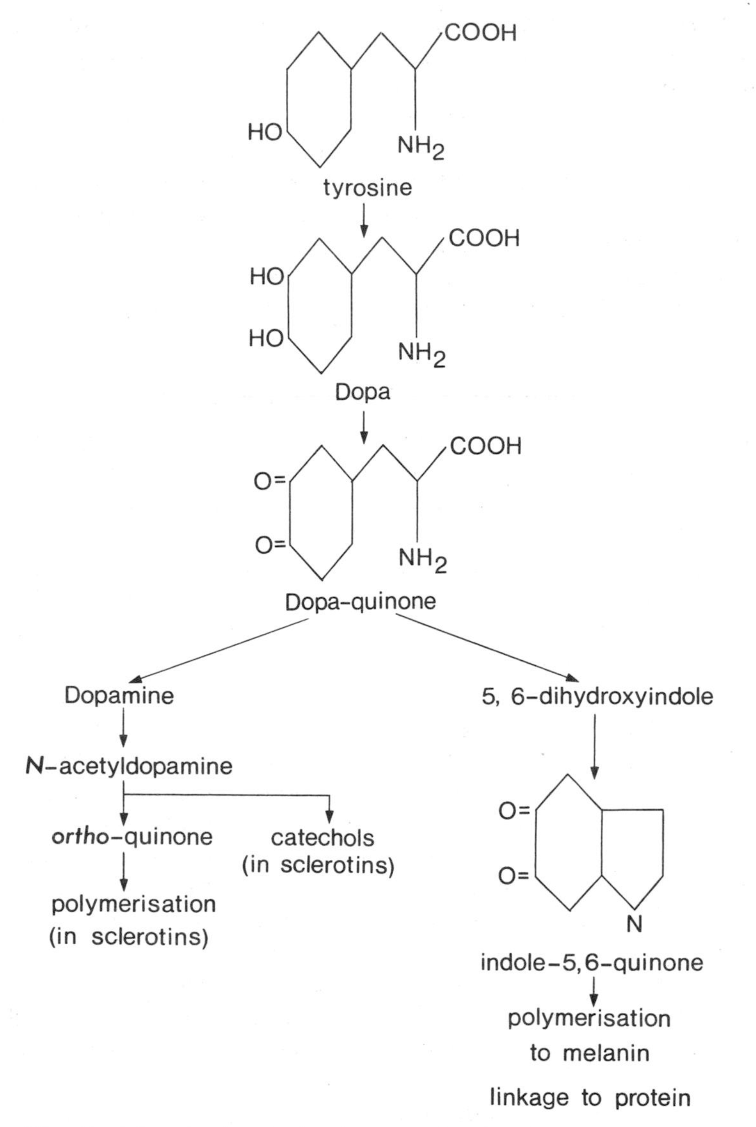

Fig. 41. Sclerotisation and melanogenesis: chemical pathways.

melanosome, probably to cysteine SH groups. Melanosomes first appear as premelanosomes derived from the Golgi apparatus, and the enzyme tyrosinase is added at this stage. The oxidation of tyrosine goes through the stages of: Dopa, Dopaquinone, leucodopachrome, Dopachrome, 5,6-dihydroxyindole, and indole-5,6-quinone. Tyrosinase catalyses the oxidations in the first two steps to Dopa and Dopa-quinone, and possibly

also later steps in melanogenesis. Tyrosinase activity is present in the unpigmented premelanosomes, but is lost once melanin polymerisation is completed.

Fully pigmented melanosomes remain in melanophores, and the melanosomes are also phagocytosed and retained by vertebrate epidermal basal cells. The presence of premelanosomes in electron micrographs (Breathnach 1971*a*, *b*) and of tyrosinase shown by formation of dark pigment when the living tissue is incubated with Dopa, are the two criteria for identification of melanocytes from macrophages and other cells (Fitzpatrick, Masamitsu and Ishikawa 1967; Riley 1972).

VARIOUS POLYSACCHARIDES

Small amounts of glycogen occur in epidermal cells, broken down to glucose or lactic acid in respiratory processes. Cellulose is found in the tunicate test and in small amounts in vertebrate dermis. Chitin is a condensation product of *N*-acetyl glucosamine with a smaller quantity of glucosamine, just as cellulose is a condensation product of glucose. It differs from cellulose in that the OH groups on the C atoms are replaced by $-NHCOCH_3$. In the older terminology now discarded by biochemists but still used by histochemists, chitin is a neutral mucopolysaccharide (Pearse 1968).

In chitin, which is an exclusively invertebrate epidermal exosecretory product, parallel chains of acetyl glucosamine and glucosamine molecules are linked together by hydrogen bonds to give microscopic fibrils. Chitin, which has a high tensile strength, always occurs in nature bound to protein and it is extremely difficult to separate as a pure substance. In the insect cuticle, chitin is linked to protein by covalent bonds. The molecular spacings along the protein closely fit those in the chitin which gives a tough lattice structure which may be further strengthened by quinone linkages to give a chitoscleroprotein. Arthropod chitin is in the alpha-helical form, but it is beta-pleated in lower invertebrates. Specific chitinases break down particular chitins but not others, indicating subtle differences in structure not yet determined by chemical means (Richards 1951; Rockstein 1964).

Hyaluronic acid, chondroitin-4-sulphate, and chondroitin-6-sulphate are associated with collagen in the dermis and occur in collagenous cuticles, while heparin is formed in mast cells. These substances contain an amino sugar such as glucosamine or glucuronic acid, and in the old terminology are termed acid mucopolysaccharides. Various other mucopolysaccharides occur in invertebrate and fish cuticles and in the lateral line cupula jelly (Hunt 1970).

Protein complexes with a variety of polysaccharides occur in the body. These include chitoproteins and protein complexes with connective tissue

mucopolysaccharides. The glycoproteins (in which the protein is larger than the carbohydrate component, the reverse of mucopolysaccharides) occur in glandular mucins. Mucopolysaccharides also occur in glandular secretions. Histochemical methods for mucopolysaccharides are less satisfactory than biochemical procedures, which is why the old terminology is still used.

SILK

Arthropod integumental glands produce a variety of chemically different types of filamentous silks. In the silkworm larva, *Bombyx*, the silk is a fibroin, but collagen-like proteins and chitin occur in silks of some arthropods. When extruded, *Bombyx* silk contains 75 per cent fibroin and 25 per cent sercin (silk gelatine). The latter is a water-soluble adhesive which makes the thread sticky, and is removed in commercial processing to reveal two cross-bonded fibroin filaments (Rudall 1963; Peakall 1969; Friedrich and Langer 1969). Silk fibroin is rich in glycine, alanine and tyrosine, and the polypeptide chains are linked by hydrogen bonds (Fig. 40*b*). X-ray analysis shows beta-pleated regions connected by randomly orientated segments. Sercin is rich in the hydroxy-amino acid serine. In the silk gland these two components are secreted in soluble colloidal form. Solidification and the appearance of a crystalline X-ray pattern occur immediately the silk is extruded, but setting is not due to exposure to air or to drying. Silk fibres show both good tensile strength and elasticity. A similar method of fibroin secretion probably occurs in the cuticle.

SOME OTHER PROTEINS

COLLAGEN

This is a fibrous protein with an alpha-helical structure which in mammals contains characteristically large amounts (one-third) of the amino acids hydroxyproline and hydroxylysine, one-third glycine, less than in fibroins, and one-third polar amino acids, but no cystine. Only minor differences in molecular structure occur in different animal collagens. Invertebrate collagens have more acid amino acids and more serine and threonine than vertebrates, while there are more stabilising (bonding) amino acids in mammalian collagens. Collagen is exosecreted by fibroblasts in the form of precursor soluble tropocollagen particles, which do not vary much in a wide variety of animals. These are linked together by ester and aldehyde groups. Each particle has a molecular weight of about 350,000 and contains three polypeptides, two similar and one different, each with about 1000 amino acids, coiled and supercoiled in a triple helix. Collagen has a characteristic X-ray diffraction pattern quite different to silk and keratin (Gross 1961; Ramachandran and Gould 1967; Harkness 1961).

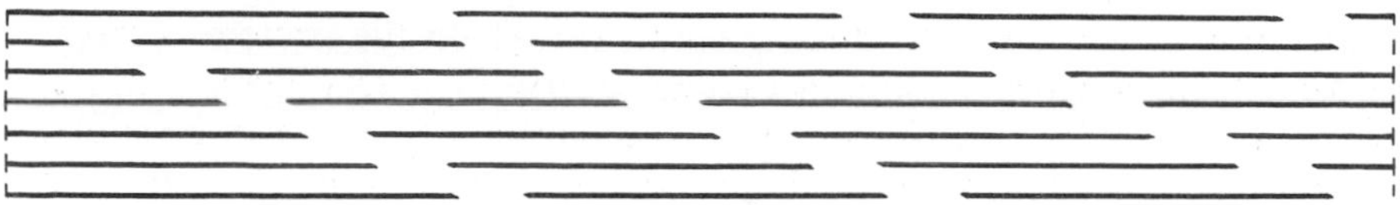

Fig. 42. Arrangement of tropocollagen with normal overlap in collagen microfibrils.

Aggregation of tropocollagen occurs in the tissue space outside the cells to form microfibrils (Fig. 42) in which the adjacent particles normally overlap (Gross 1961). Identical changes occur *in vitro* on warming a colloidal solution of tropocollagen, and artifically produced fibres can also be formed with the particles abutting. In high-resolution electron micrographs, the stained tissue collagen fibrils show characteristic cross-bands 700 Å apart, due to tropocollagen overlapping. The fine reticular network in the basal lamina is a form of collagen. Acid mucopolysaccharides are closely associated with collagen and play an important part in aggregation.

Cuticular collagen is secreted by the epidermal cells of certain invertebrates, and also it is secreted generally by the epidermis into the basal lamina. The enzyme collagenase hydrolyses all collagens.

ELASTIN

This elastic protein occurs in vertebrate skin connective tissue and is seen as a network of optically visible filaments even in fresh tissue. Elastin has not been shown in invertebrates although related proteins may occur. It resembles collagen in chemical composition, but contains more neutral amino acids and less glutamic acid, aspartic acid, arginine and lysine. The rare amino acid desmosine forms cross-linkages not found in collagen. Some phospholipid is bound to elastin. In its molecular structure, elastin is quite different to collagen; it is much more stable, and even resists boiling water.

KERATIN

This differs from the proteins just mentioned in being an endocellular product typical of vertebrate epidermis (Alexander, Hudson and Earland 1963; Fraser 1969; Spearman 1966). It has a much more complex quaternary molecular structure than silk, collagen or arthropod sclerotins, and in consequence it is most unlikely to occur as an exosecretion as has been suggested in certain invertebrates.

Although not peculiar to keratin, an important constituent is the di-amino acid cystine, formed by exergonic oxidation with release of energy from two molecules of the sulphydryl amino acid cysteine in neighbouring

chains, or different points on the same chain. The SS cross-linkages render the protein resistant to proteases, but hydrogen bonds, salt linkages and Van de Waals forces are also important linkages (Fig. 40*a*). In addition to cystine and cysteine, major amino acids present include glutamic acid, arginine, leucine, glycine and tyrosine. Wool keratin contains some eighteen different amino acids.

Keratins typically have a two-component structure. One component is a fibrous protein which in mammalian hairs is made up of alpha-helical chains with few cystine bonds, complexly coiled to give rope-like filaments alternating with randomly orientated segments. In feathers, the filaments contain beta-pleated chains which are richer in cystine than alpha filaments. The second component is the globular matrix protein which lies between the filaments. One type of globular protein is rich in cystine and another type is characterised by glycine and tyrosine, but has little cystine. Both globular proteins have an amorphous X-ray pattern. The different components, which can be separated by specific bond-breaking agents, contain a variety of polypeptides shown by their N-terminal amino acids, electrophoretic movements, and different molecular sizes (Gillespie 1970). The cystine-rich fraction in sheep's wool is increased on a cystine-rich diet, although only within strict limits. In view of the number of its different polypeptide components, at least fourteen different structural genes possibly in different chromosomes are involved in elaboration of sheep's wool keratin.

SUMMARY

Elaboration of polypeptides occurs in the ribosomes coded by messenger RNA. Their architectural arrangement in alpha and beta chains, or coiled as globular protein, is determined mechanically by the distribution of amino acids able to form cross-linkages. Several different polypeptides may link together to form high molecular weight proteins. The most complex in the integument is keratin which in consequence can probably be assembled only in the confined space of a cell. Collagen is formed by aggregation of similar particles.

Cuticular proteins include fibroins of arthropods and collagen of nematodes and annelids. These are sometimes tanned by interpolation of quinone polymers between the polypeptide chains. Quinone tanning and melanogenesis are similar processes and melanin occurs both inside and outside cells.

Invertebrate cuticles are sometimes strengthened by chitin fibrils, and other mucopolysaccharides are often present in the cuticle as well as in the dermis.

16

TRANSPORT THROUGH THE SKIN

The unimpeded inward or outward movement of water and ions through the integument must upset the salt balance of living cells, while the leakage of proteins from tissue fluids imperils life. Different groups of animals have tackled the problem or come to terms with skin permeability in a variety of ways, with least success in marine invertebrates and most efficiently in insects and terrestrial vertebrates (Davson 1970).

INVERTEBRATES

MARINE SPECIES

These all have tissue fluids isotonic with sea water so that osmotic movement is eliminated and only diffusion occurs. Septate desmosomes (Fig. 3) between the outermost epidermal cells occur in most invertebrates with a cellular epidermis and prevent diffusion of water and salts through the intercellular spaces. Others with a syncytial epidermis have the intercellular pathway eliminated. Passage is therefore mainly through the epidermal lipoproteinous plasma membranes, which vary in permeability to ions but are impermeable to tissue proteins. This membrane barrier enables mammalian red blood cells to differ in ionic composition from the surrounding plasma. In marine invertebrates the outward-facing cell membranes usually constitute the physiological interface between the sea and tissue fluid even when a thin cuticle is present.

Most soft-bodied invertebrates, such as molluscs, change slowly in volume when placed in solutions of different osmolarity. Diffusion of radio-active labelled substances in echinoderms shows that the skin is more permeable to ions and small molecules than to larger molecules.

The presence of a cuticle or a layer of mucin is little hindrance to penetration, except in thick cuticles of decapod crustaceans. In the lobster *Homarus*, the skin is more permeable during and just after ecdysis, but becomes much less so when the carapace has hardened. Permeability to water is probably reduced by the presence of protein-bound lipoidal substances and waxes. However, even here water movement is influenced by a diuretic hormone from the sinus gland which affects epithelial cell salt permeability.

BRACKISH-WATER SPECIES

River estuaries and tidal pools are liable to rapid changes in salinity, and animals in these habitats must contend with a wide range of osmolarities. Most brackish-water invertebrates have come to terms with this problem by being able to withstand a wide range of tissue salt concentrations. When the habitat salinity is lowered, water enters through the skin by osmosis and dilutes the body fluids, and conversely water is sucked out when the habitat salinity is increased. Even so, dilution of body fluids by a factor of two is the maximum tolerated. A few species, such as the shore crab *Carcinus* and the polychaete worm *Nereis*, have some power of osmoregulation which involves active transport of ions against the osmotic gradient, discussed in the next section.

FRESH-WATER SPECIES

These animals have to contend with osmotic movement of water through the skin into the animal because of the greater osmolarity of the tissues. The excess water taken in is removed by excretory organs, or in fresh-water protozoa by contractile vacuoles, and involves retention of ions. The loss through the epidermis of sodium and chloride ions is prevented in most fresh-water invertebrates by active transport against the diffusion gradient, involving energy expenditure. The mechanism is similar to that described later in amphibian skin and is not found in marine invertebrates except in a few brackish-water species.

LOWER VERTEBRATES

CYCLOSTOMES AND FISH

Marine species

The stratified epidermis of cyclostomes and fish is less permeable to ions than the thin invertebrate integument. Hagfish blood is ionically isotonic with sea water, so that osmotic movement is eliminated, but marine bony fish are hypotonic to sea water, and so water is lost through the skin and gills by osmosis. This is made good by drinking sea water, and the excess salt is removed from the blood by active transport across specialised cells in the gills. Kidneys are not important in salt regulation in marine teleosts. Elasmobranchs keep their blood isotonic with sea water by means of urea.

Fresh-water species

In these, the blood is hypertonic to the surrounding water which is absorbed through the skin and gills. Fresh-water fish do not drink and the excess water absorbed is excreted by the kidneys as hypotonic urine. Loss

of salt into the surrounding water appears unimportant in the skin although active transport does not occur in fish epidermis, for surprisingly it is almost impermeable to ions unless injured. Active transport of salt against the osmotic gradient in the gills prevents leakage into the surrounding water.

AMPHIBIA

Most Amphibia live in fresh water and so sodium and chloride ions tend to move from the hypertonic blood out of the body and water enters through the skin. Movement of water into or out of the body between the epidermal cells is probably prevented by fused lateral junctions between the outermost cells (found in other vertebrate epithelia showing active transport). Transport of water is therefore directly across the cell membranes (Farquhar and Palade 1965). Recently, fused junctions between superficial epidermal cells have been found in many vertebrates which do not show active transport. The horny layer, when this occurs, probably slows down passage of water because of its protein-bound lipids, but is more an adaptation to reduce water loss by evaporation, and many aquatic species are unkeratinised.

Salt is prevented from leaving amphibian skin by active transport against the diffusion gradient by means of 'pumps' in the living epidermal cells (Fig. 26). The plasma membrane-bound enzyme, ATPase, breaks down ATP to lower phosphate compounds, and the energy released is used for active transport of both sodium and chloride ions, moved independently. Diffusion is passive into the epidermal cells from the extracellular fluid, for the plasma membrane is permeable to inward movement, but impermeable to outward movement of labelled sodium. Once in the cytoplasm, ions are probably bound to a lipoidal carrier molecule, and the complex then passes back across the plasma membrane, most likely in specific pumping sites which contain membrane-bound ATPase and by means of a specific permease for the carrier molecule. Permease enzymes actively transport organic molecules across cell membranes against diffusion gradients. Phosphatidic acid, which has an affinity for sodium, is probably one carrier molecule since its turnover is increased in the avian salt gland during active salt secretion. A net flux of ions towards the blood is achieved if the pumps are restricted to inward-facing sites with fused junctions at the outside. A local high concentration of ions pumped into the epidermal intercellular space reverses the diffusion gradient so that the flow is towards the dermis. Salt retention in this way indirectly promotes the passive flow of water into the skin by osmosis and excess water is removed by the kidneys in hypotonic urine. Other ions, such as calcium and ammonia, move by simple diffusion through the epidermis. Active

transport of sodium and chloride is stopped by dinitrophenol which does not interfere with oxygen consumption but prevents synthesis of ATP. This suggests that oxidative processes are not involved. Indeed, in other tissues, such as the mammalian intestinal mucosa, active transport is associated with the presence of membrane-bound ATPase.

The posterior pituitary lobe antidiuretic hormone, ADH, when injected into frogs and toads stimulates the active transport of water by the epidermal cells. Aldosterone probably promotes influx of ions. In the mammalian kidney tubule this hormone promotes salt resorption. Probably in frogs and toads ADH is released in response to impulses from the brain osmoregulating centre, but not all Amphibia respond in this way.

Two species of frogs which inhabit brackish water have got over osmosis in different ways. One has an elevated blood salt level, as in the hagfish, and the other a high blood urea level, as in elasmobranchs; both of which make the blood isotonic with the surrounding water. Active transport does not occur in higher vertebrate epidermis (Davson 1970; Potts and Parry 1964; Prosser and Brown 1961).

LAND ANIMALS

Animals which live in contact with the atmosphere have to contend with quite different problems to aquatic species. Osmosis, which is the main problem for life in water, is replaced in land animals by evaporative water loss from the body surface (transpiration).

INSECTS

The first animals which overcame the problem of transpiration, and indeed the most successful, were insects. Other arthropods, notably spiders and myriapods, almost succeeded but were held back by a less efficient mechanism for atmospheric breathing than the insect tracheal system. The insect skin, although permeable to organic insecticides, has developed a highly waterproof cuticle to transpiration and slightly less so to water absorption achieved by the outer monolayer of hydrophobic wax molecules in the epicuticle. These waxes are esters of even-numbered primary alcohols with fatty acids. Ketonic alcohols occur in several insect waxes. Bee's wax is an example of a similar wax produced as a bulk glandular secretion. Although waxes and lipids also occur lower down in the cuticle, these are less important in reducing permeability. Thickness of the cuticle is therefore no guide to its permeability and some of the most impervious insect cuticles are quite thin. Heating the cuticle to 60 °C greatly increases its permeability to water, probably by melting the wax monolayer, and it can also be removed by solvents. A hydrophobic wax layer also occurs in

the leaf cuticles of plants but does not prevent water absorption, a fact utilised in foliage feeding with nutrients.

The cuticular barrier, together with septate desmosomes, control of spiracle openings, water retention in the gut and excretory organs, and conservation of metabolically-formed water, has enabled insects to thrive under desiccating conditions (Ebeling 1964).

VERTEBRATES

In land vertebrates, the skin is passively waterproofed by the covering layer of dead keratinised cells. Keratin is hygroscopic and takes up water in equilibrium with atmospheric humidity, so that waterproofing is probably achieved by protein-bound phospholipids. The keratin phospholipid complex thus behaves as a molecular sieve allowing water to pass through the horny layer in either direction through the keratin, but not through the phospholipid (Spearman 1970*b*). It is in consequence a less efficient barrier than the insect cuticle. The lateral junctions between the transitional cells appear fused in higher vertebrates and are relatively impervious to small molecules.

In terrestrial Amphibia, because the skin is a respiratory organ, the horny layer is too thin to provide adequate protection against desiccation and water is also lost in glandular secretions. A few amphibians contend with desert conditions, achieved by a thickened horny layer, a less glandular skin, and water storage in the urinary bladder.

The dry, almost non-glandular skin of reptiles with their thick phospholipid-rich horny covering is much less permeable to water than in amphibians. Nevertheless, there are wide species differences in rates of reptilian transpiration (Bellairs 1969). The compact reptilian surface layer is repellent to rain droplets (Bentley and Schmidt-Nielsen 1966).

HIGHER VERTEBRATES

TERRESTRIAL BIRDS AND MAMMALS

Birds have a thin horny layer, rich in keratin-bound phospholipids, as in reptiles. Water penetration through the plumage in aquatic species is reduced by hydrophobic lipids from the preen gland.

Typically in terrestrial mammals, protein-bound phospholipid is retained in the horny cells only in a thin compact zone in the base of the stratum corneum, referred to as the s.c. conjunctum, and above this region the phospholipid is lost from the horny cells by continued autolysis (Spearman 1970*b*). In excised human skin, the horny layer is least permeable to water penetration in the conjunctum region (Tregear 1966; Malkinson 1964). Dissolved sebum is not important for waterproofing the

skin since the lipoidal constituents are water-miscible, but in aquatic mammals, such as beavers and seals, sebum may contain hydrophobic lipids since the fur does not get wet. Mammalian skin is not very waterproof and losses must continually be made good by drinking. The amount of water lost by skin transpiration in man after sweat glands have been inhibited by atropine amounts to 120 ml of water per square metre of exposed skin surface in 24 hours under comfortable conditions of temperature and humidity. The pelts of terrestrial mammals provide little protection against water loss, but hairs have water repellent surfaces against rain. The skins of laboratory mammals are some ten times more permeable to water than human skin.

AQUATIC MAMMALS

A thick, bound, phospholipid-rich horny layer occurs in seals, whales, manatees and the hippopotamus (Spearman 1966, 1970*b*), and probably increases the barrier to water penetration. In the absence of active transport, the only inhibitory effect on ionic movement through the epidermal intercellular pathway is the tendency of cations to bind to negatively charged membranes.

In terrestrial mammals, including man, considerable amounts of water are absorbed through the skin when the body is immersed for long periods.

PARASITIC WORMS

In tape worms, flukes, and the Acanthocephala, the skin is a food-absorbent organ with absorbent microvilli, permeable to amino acids and simple sugars, unlike the integuments of other animals.

CIRCULATION OF SUBSTANCES IN THE SKIN

Once nutrients, hormones and oxygen have left the blood capillaries, the dermis and basal lamina present no barrier to diffusion. For substances which have entered between the epidermal cells, there is a continuous intercellular pathway which in vertebrates extends from the basal layer to the transitional layer. Epidermal cells readily take in substances by pinocytosis, a form of ultrastructural phagocytosis. Substances withdrawn from the tissue fluid pass through the plasma membrane into the cytoplasm to form microvesicles. Hydrophobic organic substances absorbed through the skin are conjugated with protein in the cells which makes them water-miscible for transport in the blood as chylomicrons. Many such potentially toxic substances are met with today, products of civilisation.

Diffusion back to the blood of metabolites and carbon dioxide also follows the epidermal intercellular pathway which, in effect, is an extension of the dermal intercellular space. The skin is a respiratory surface in invertebrates with a thin cuticle, in fish and Amphibia, but is not important in higher vertebrates except for superficial cells.

SUMMARY

Marine invertebrates are isotonic to sea water but fresh-water species have hypertonic body fluids and absorb water by osmosis, later removed by the excretory organs. Loss of salt from body fluid in fresh-water species is prevented by active transport in the epidermis.

Marine fish, except for the hagfish and elasmobranchs which are isotonic to sea water, have hypotonic blood and lose water by osmosis.

Fresh-water fish are hypertonic and so gain water. Active transport in the gills prevents salt loss. Fish epidermis is much less permeable than invertebrate skin.

Amphibia show active transport of salt in the epidermis, but this does not occur in higher vertebrate skin. Here the water barrier is passive in the horny layer and consists of hydrophobic phospholipids linked to keratin.

The insect cuticle is waterproofed by a layer of wax in the epicuticle.

The epidermal intercellular space is continuous with the dermal space and is the normal pathway for diffusion of nutrients and metabolites.

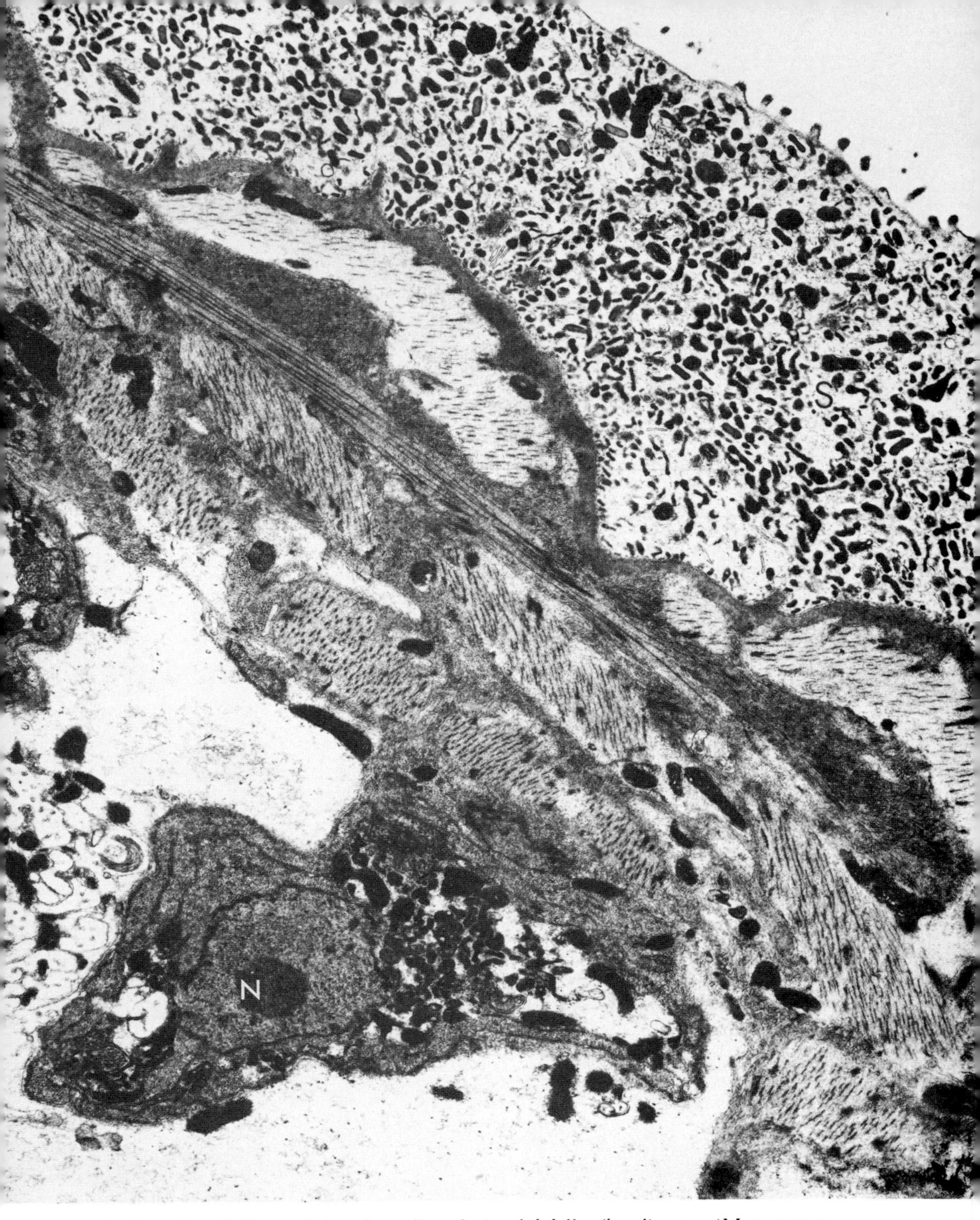

1. Epidermis of blood living juvenile of *Amphibdella flavolineata* (Monogenea Trematoda) showing the epidermal surface syncytium, S, with absorbent microvilli and one insunken epidermal cell body with nucleus, N. Electron micrograph $\times$ 29,250. (Courtesy: K. M. Lyons, *Parasitology* **63**, 181.)

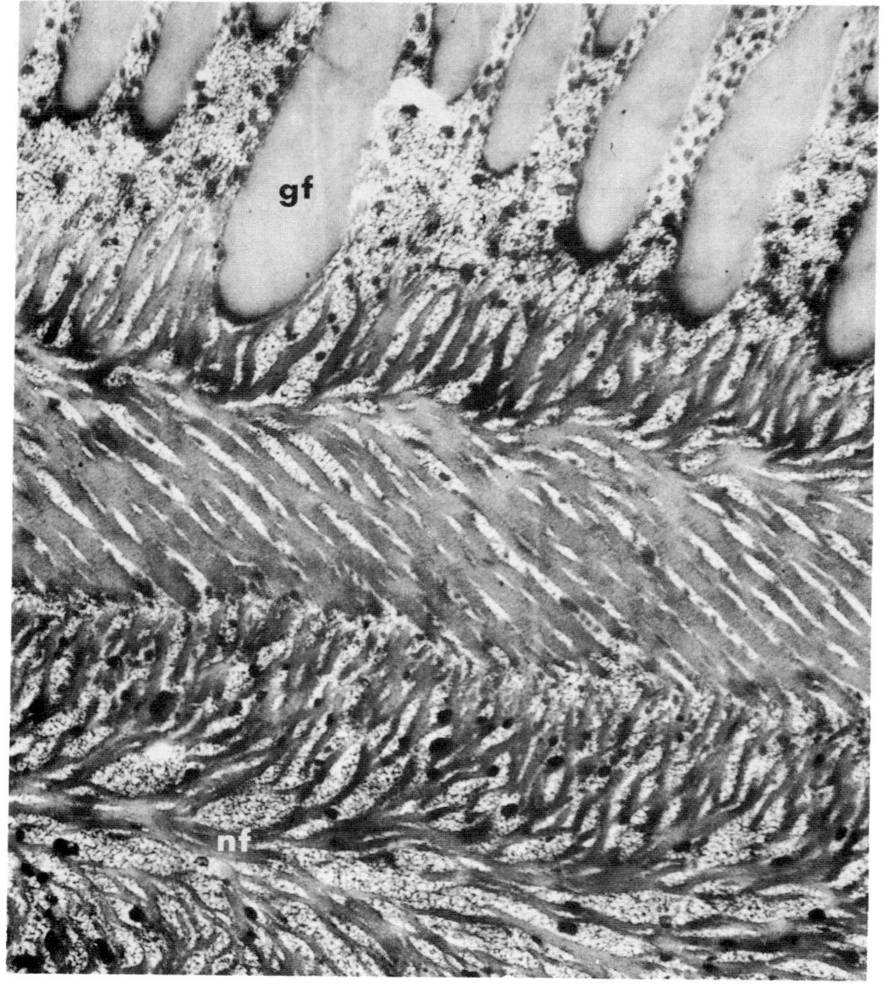

2. Vertical section through part of the cuticle of a nematode, *Mermis nigrescens*, showing giant fibres, gf; network of fibres, nf. Both are composed of collagen. Electron micrograph $\times$ 52,000. (Courtesy: D. L. Lee, *J. Zool., Lond.* **161**, 513.)

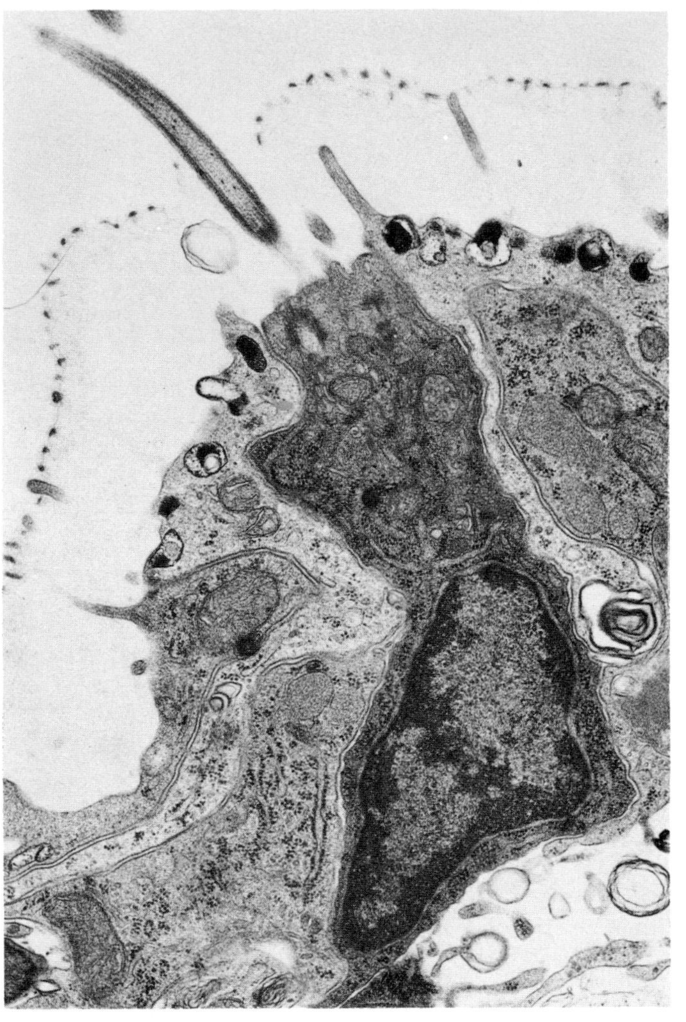

3. Epidermis of *Aeolosoma*, a fresh water oligochaete (Annelida). A few ciliated cells occur under the prostomium. One of these is shown with adjacent cuticularcytes. Microvilli extend to the cuticular surface. Electron micrograph. (Courtesy: H. E. Potswald, *J. Morph.* **135**, 185.)

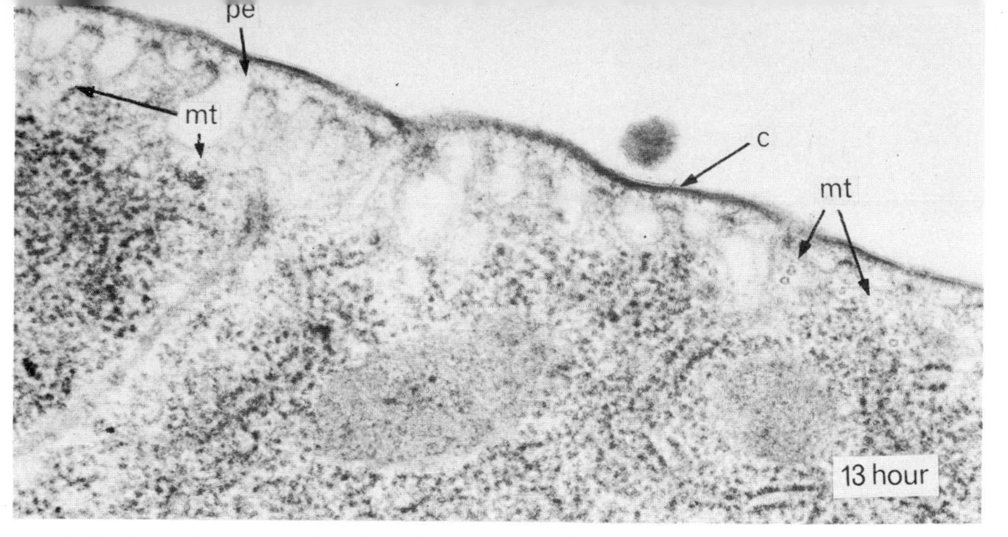

4. Early cuticle formation in embryo of the fruit fly *Drosophila melanogaster*. Secretory edge of epidermal cells. The microvilli containing microtubules, mt, are covered by secreted cuticulin, c. Later the protein epicuticle, pe, and endocuticle are laid down under the cuticulin layer. Electron micrograph ×47,000. (Courtesy: R. Hillman and L. H. Leenik, *J. Morph.* **131**, 383.)

5. Electron micrograph of calcified repair layer within one hour of removal of a piece of shell from the land snail *Helix pomatia*. Both large and small crystals are deposited; the largest apparently associated with electron dense areas, probably organic material. ×23,000. (Reproduced by permission of the National Research Council of Canada from K. M. Wilbur and A. S. M. Saleuddin, *Canad. J. Zool.* **47**, 51 (1969).)

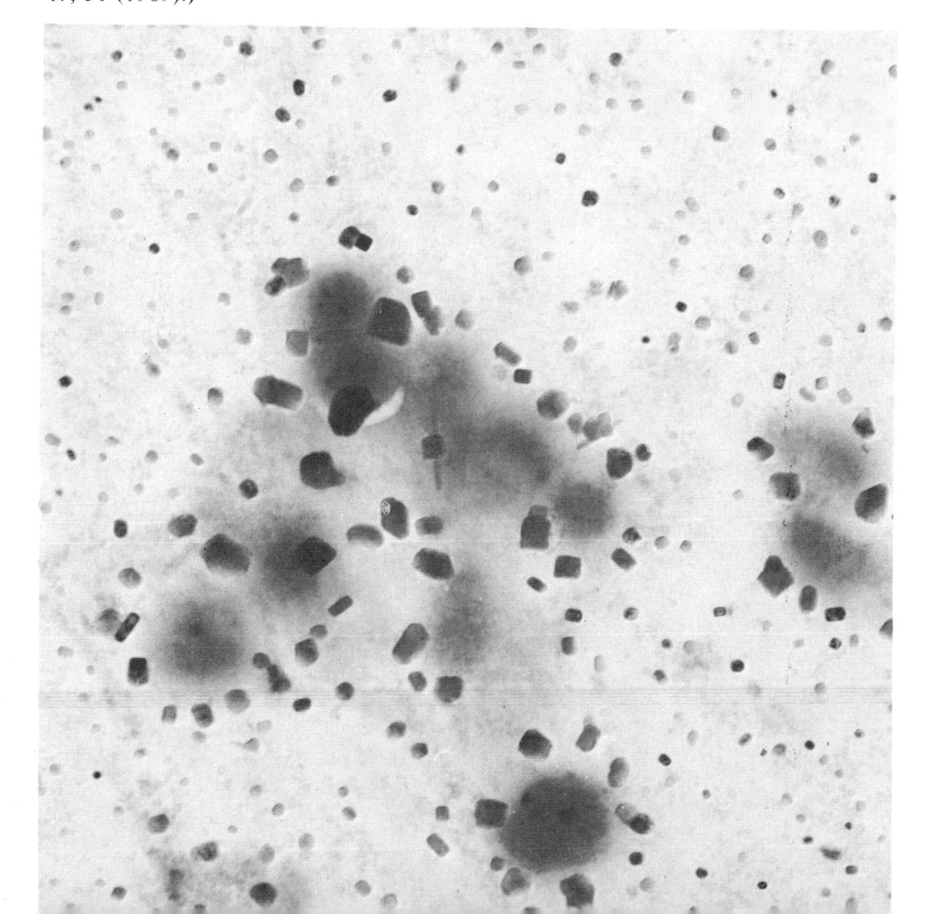

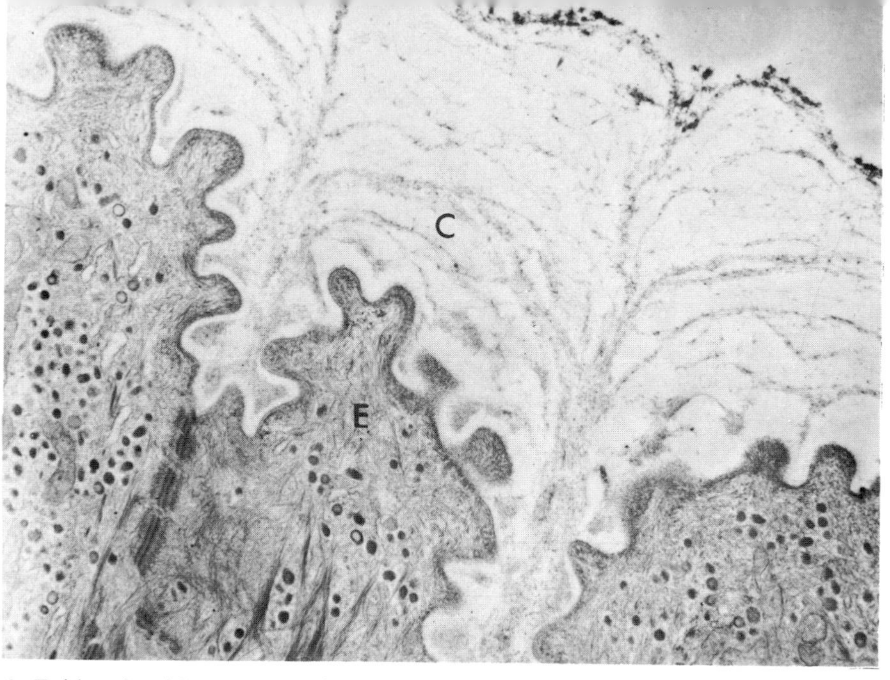

6. Epidermis with secretory microvilli, E, and cuticle, C, from pectoral fin of the teleost fish *Trigla*. Electron micrograph stained with PTA. (Courtesy: M. Whitear, *J. Zool., Lond.* **160**, 437.)

7. Parakeratotic breeding tubercle from head of the teleost fish *Leuciscus*. Epidermis, e; horny layer, k; basal lamina, bm. H. & E. stain × 106. (Courtesy of the American Museum of Natural History; W. L. Wiley and B. B. Collette, *Bull. Am. Mus. Nat. Hist.* **143**, 147.)

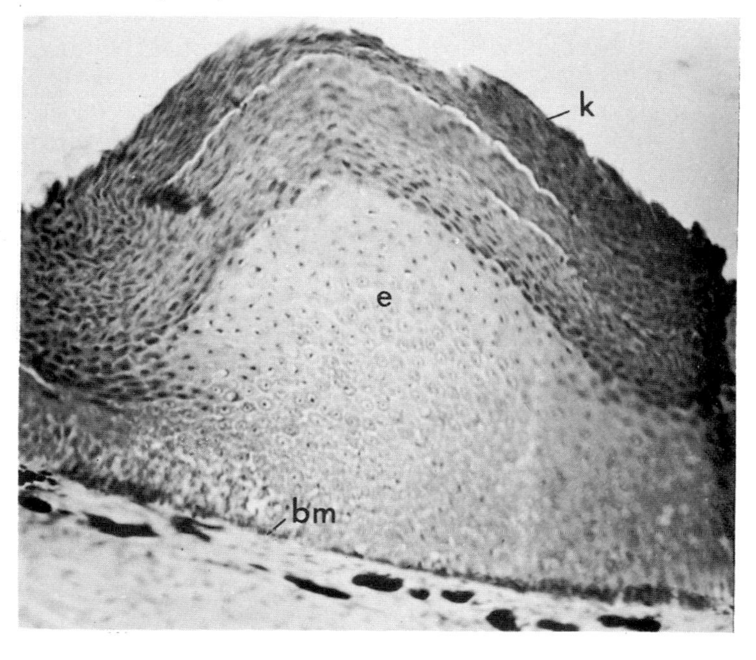

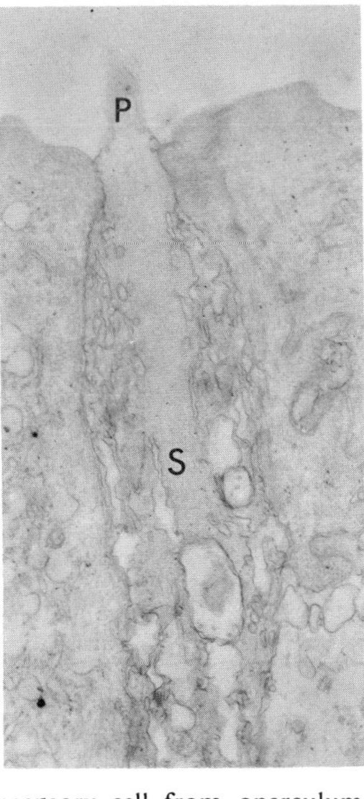

8. Vertical section chemosensory cell from operculum epidermis of the teleost fish *Pomatochiatus*. Sensory cell, S; sensory process, P. Electron micrograph stained with lead citrate. (Courtesy: M. Whitear, *J. Zool., Lond.* **163**, 237.)

9. Vertical section moulted parakeratotic horny layer of frog *Rana temporaria* with attached flask cells. H. & E. stain. (R. I. C. Spearman, *J. Morph.* **125**, 129.)

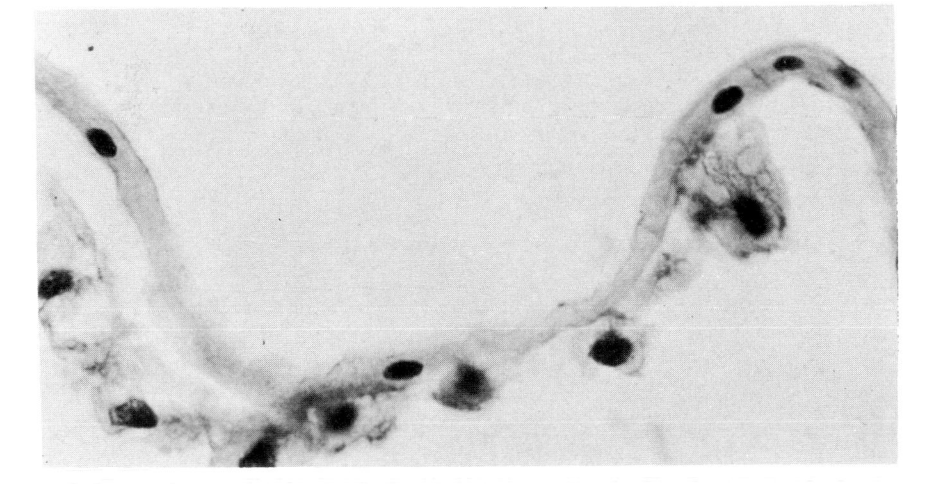

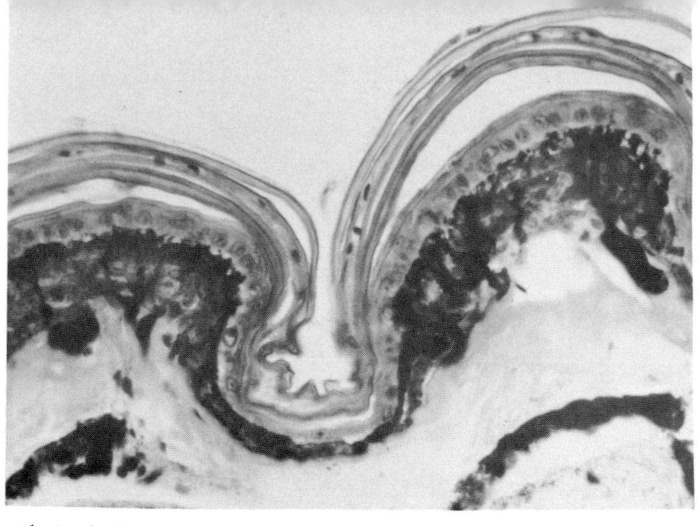

10. Lizard *Anolis* back skin. Two horny layer generations are separated by an unkeratinised layer of nucleated clear cells. The space between the lowermost horny layer and the epidermis underneath is artifactual. Large melanophores are shown immediately under the epidermis. H. & E. stain. (Courtesy: R. I. C. Spearman (1966), *Biol. Rev.* **41**, 59.)

11. Melanocytes from black cavy ear skin in tissue culture on glass. × 1600. (Courtesy: P. A. Riley, unpublished.)

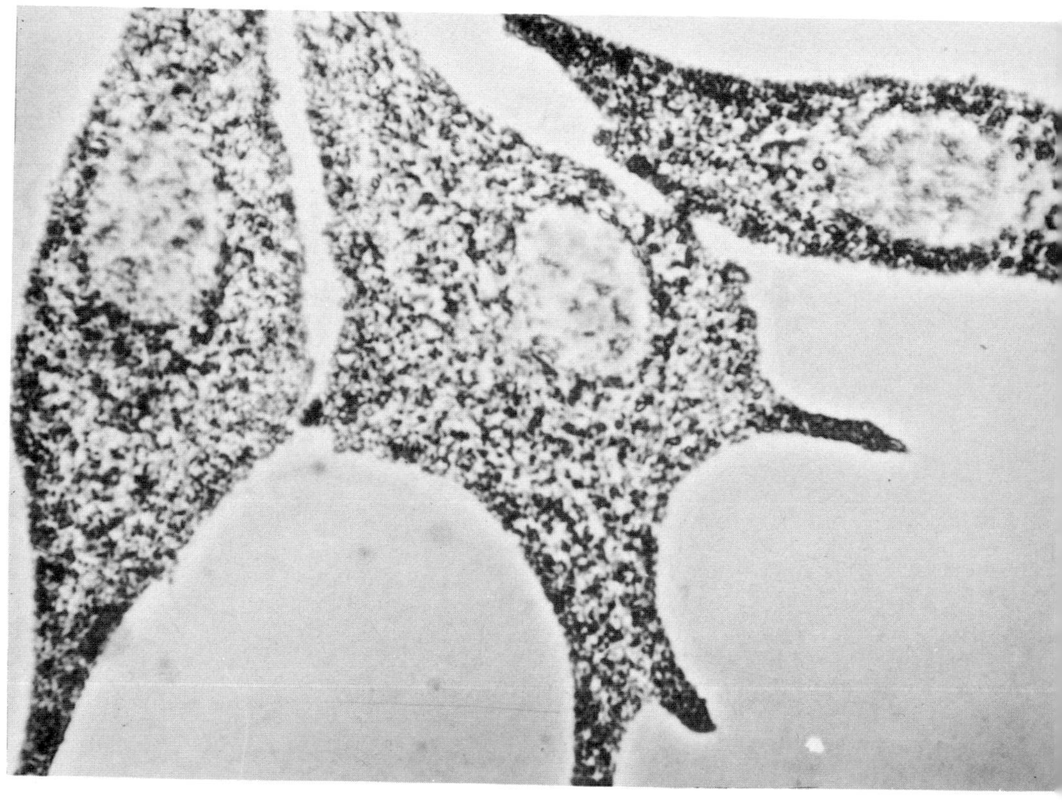

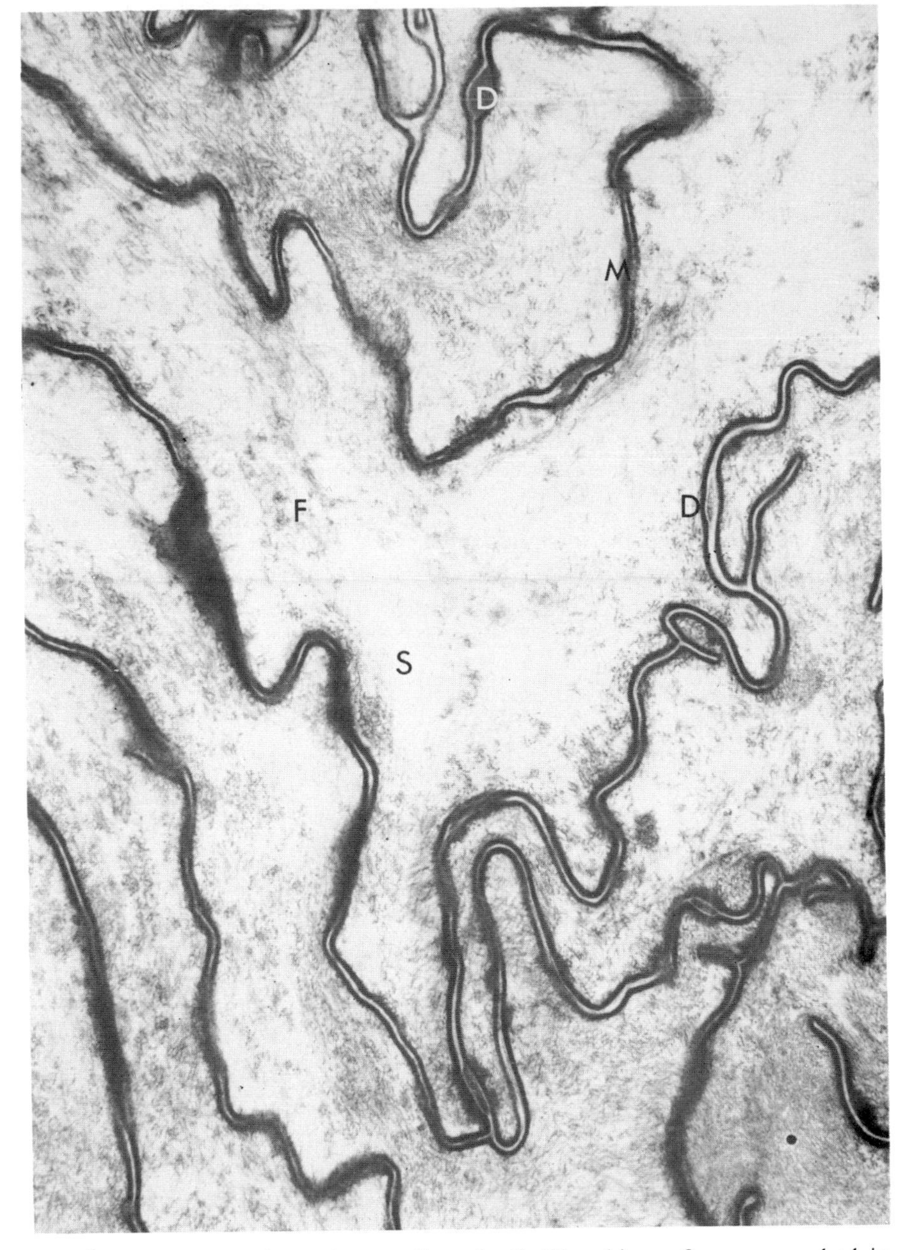

12. Cavy back horny layer: intermediate depth. The skin surface was washed in water and ethanol prior to excision to remove soluble breakdown products from the keratinised cells, and was then fixed in buffered osmium tetroxide and stained with lead citrate. Keratin fibres, F; intracellular space, S; intercellular space with desmosomes, D. The horny cell membranes, M, are deeply osmophilic and have a thin cytoplasmic shell. Electron micrograph ×30,000. (Courtesy: R. I. C. Spearman, unpublished.)

17

IMMUNITY

TYPES OF IMMUNE RESPONSE

The immune response is the mechanism whereby an animal destroys certain foreign substances which enter the body. Invasive agents which elicit such a response are termed antigens. Primarily, immunity is a defensive mechanism, although sometimes in higher animals the host tissues are also damaged. Three types of immune response occur. The most primitive is phagocytosis, helped by the foreign material being coated with opsonin, an unspecific immunoglobulin. More advanced are humoral antibodies (specific immunoglobulins) which are liberated into the blood and react with circulating antigens such as in bacteria and foreign red blood cells. In another type of reaction, antibodies are retained inside lymphocytes which migrate to the site of the antigen, a method used to destroy transplanted foreign tissue.

NATURE OF ANTIBODIES

These are complex high molecular weight proteins in the gammaglobulin series, elaborated from the products of several genes. Probably the amino acid sequence in the protein determines specificity, and combination of products of different genes in a multitude of ways enables antibodies to be formed against such a wide variety of antigens in nature.

IMMUNITY IN DIFFERENT CLASSES OF ANIMALS

Invertebrates do not show a highly organised immunity. Turbellaria show no immune response and incorporate coelenterate nematocysts into their tissues. In the more advanced groups such as insects, phagocytes congregate around and attack foreign bodies, but tissues can still be transplanted readily within the same species. Nevertheless, transplants between species (xenografts) are destroyed, and the fact that a second injection of some foreign material induces a more rapid reaction suggests an opsonin-type mechanism.

Cyclostomes react weakly to antigens, but elasmobranchs and teleosts show homograft rejection (between different members of the same species),

and a highly developed immunity occurs in higher animals. In cold-blooded animals, the rejection mechanism is slowed down by cold ambient conditions, and even in fish it works better at around 37 °C, mammalian blood temperature.

HUMORAL ANTIBODIES

These play little part in the skin and so the mechanism will be outlined briefly. Although some antibodies are present in the circulation even without contact with antigen (for example blood groups), immunity is generally achieved by an antigen triggering off the formation of a specific antibody. In mammals, circulating bone marrow-derived 'B' lymphocytes (derived in birds from a lymphoid gland, the bursa of Fabricius) are primed by contact with the antigen, so that on return to the lymph nodes they divide to form a store of memory cells. A later invasion by the same antigen, detected by circulating 'B' lymphocytes, causes proliferation of memory cells in the nodes which give rise to plasma cells, which synthesise and release into the circulation specific antibody.

Another type of mononuclear cell, the thymus-derived 'T' lymphocyte, also in the lymph nodes, helps the 'B' cells in many types of reaction by making them more sensitive to antigen.

DELAYED HYPERSENSITIVITY

This is due to the direct action of 'T' lymphocytes on antigenic material, probably aided by 'B' lymphocytes. Foreign tissue grafts are destroyed in this way, and these cells are also responsible for reactions of the delayed hypersensitivity type in which local areas of damage or death of skin cells occur. This type of reaction is quite distinct from inherited allergic reactions (atopies) which are immediate and are caused by a weak type of circulating antibody (reagin) formed only in these subjects, and which stick to the plasma membranes so that tissues are damaged by the subsequent allergen–reagin reaction.

HAPTEN REACTIONS

Delayed hypersensitivity lesions are produced by haptens, a variety of chemical agents with one feature in common: after conjugation on to protein within the skin cells, or combination with tissue protein, they become powerfully antigenic. An example is the experimental sensitisation of guinea-pig skin to dinitrochlorobenzene (DNCB). When this is applied to a small area of skin, no immediate adverse reaction occurs (in

contrast to allergens), but if a later application is given to any site, a severe local reaction takes place. Histological examination of the lesion reveals an infiltrate of lymphocytes. This is a dermatitis.

HOMOGRAFT REJECTION

Autografts are from one site to another site of the same animal. A syngenic homograft is when donor and host are genetically identical, as with inbred strains of mice and in identical twins. The grafted tissue is not rejected and grows as well as an autograft. Allogenic homografts are between genetically dissimilar members of the same species, and are normally rejected after only a few days. Xenografts are even more rapidly rejected.

Rejection of grafted skin first became a problem in wartime plastic surgery, and the phenomenon was later studied in laboratory animals, mainly in mice. Following rejection, subsequent homografts from the same donor are rejected more rapidly, and it was then discovered that accelerated rejection even of primary homografts was achieved after injection of lymph node cells from another mouse of the receptor's strain which had previously been immunised to the particular donor homograft. However, transference of serum did not affect skin rejection, which indicated that humoral antibodies were not involved, although this is not true of some other tissue transplants. Moreover, lymph node cells from a mouse sensitised to a homograft, when injected into the graft donor's skin caused a local hypersensitivity reaction, but only living cells produced this effect. Clearly, therefore, both homograft rejection and hypersensitivity reactions to haptens are affected by lymphocytes, and are the same processes.

The movements and derivations of lymphocytes from the bone marrow and thymus which seed the lymph nodes have been followed by labelling the cells with H^3-thymidine, and it is clear that only recently proliferated 'T' lymphocytes are involved in homograft rejection.

Acquired tolerance to allogenic homografts can be induced by injection of donor cells into a mouse foetus. Afterwards, tissue from the donor strain is readily accepted by the recipient in adult life.

In delayed hypersensitivity, sensitisation occurs when 'T' lymphocytes come into contact with antigenic material, which may be in epidermal cells or the dermis. When later cells come in contact, one containing antigen and the other the lymphocyte containing antibody, a powerful reaction takes place at the plasma membranes, which are damaged in the process (Holbrow 1970; Roitt *et al.* 1969).

MAMMALIAN MACROPHAGES

These process antigen and make it more acceptable to 'T' lymphocytes. A substance is then secreted by 'T' lymphocytes which transforms macrophages into migrant 'killer' cells.

SUMMARY

In higher invertebrates, foreign substances are destroyed by phagocytosis, but highly developed immunity involving specific antibodies first appears in fish. Circulating antigens are destroyed by serum antibodies. Vertebrate skin grafts between genetically unrelated individuals are destroyed through a direct reaction with lymphocytes which migrate to the antigenic donor tissue and destroy it.

18

DEVELOPMENT

EMBRYONIC ORGANISATION

The fertilised egg, by cleavage and differentiation along different paths, forms the variety of adult tissues. This is accomplished in sequences of steps until finally only those genes responsible for its characteristics are switched on in a tissue. Thus, a mammalian thyroid cell has the genes for thyroxine synthesis switched on, but it is unable to form collagen. Conversely, a fibrocyte forms collagen but not thyroxine. Finer details, such as the type of epidermal development (modulation) continue to be controlled in adult skin by inductive mechanisms.

IMPORTANCE OF THE EGG CYTOPLASM

This influences the fates of early blastula cells and is most marked in the mosaic eggs of annelids, molluscs and ascidians, which show rigid cleavage patterning with different zones in the egg discernible before cleavage. In most other animals, regional differences in the cytoplasm are less obvious but are never unimportant.

POSITION OF CELLS IN THE BLASTULA

Fates of different cleavage cells are influenced by their relationships to other cells. Important factors are cell contacts and diffusion gradients of nutrients and metabolites.

THE EARLY PATTERN OF DEVELOPMENT

Prospective germ layers can be mapped out in the early frog gastrula. Ectoderm transplanted to other sites at this stage is capable of forming mesodermal or endodermal tissues. In tissue culture, isolated prospective ectoderm cells grow as a sheet of simple epithelium. Separation of the ectoderm into neural plate bordered by neural crest, in turn bordered by epidermis, is determined late in gastrulation through induction by the archenteral mesenchyme (primary organiser).

LATER REGULATION OF DEVELOPMENT

Gastrulation heralds the start of a series of inductive events. For example, towards the end of the sequence the mesodermal core of a developing denticle or tooth induces the epithelium above it to lay down enamel.

CELL MIGRATIONS

The vertebrate embryonic neural crest cells migrate and give rise to melanocytes, adrenal medullary cells, sympathetic neurones, sensory neurones of the dorsal root ganglia, and Schwann cells lining the nerves. Both melanin and adrenalin formed by these cells are chemical derivatives of Dopa. Migration of melanocytes from chick neural crest has been followed by labelling nuclei with H^3-thymidine. If labelled neural tube and crest are grafted into a non-labelled embryo in place of its own, movement of labelled cells can be followed as development proceeds. These chimeric grafts in embryos are not rejected.

The vertebrate dermis and subcutaneous tissue are derived from the lateral plate mesoderm with a contribution from the somite dermatomes. Cells from these regions migrate to line the entire underside of the epidermis, which later inductive effects suggest occurs in an ordered manner with particular cells to certain sites, such as to foot pads, hair follicle papillae, and tooth papillae. Considerable movement of mesenchymal cells occurs during morphogenesis and only ceases when the cells come in contact with one another (contact inhibition). The developing dermis, in consequence, becomes packed with cells which are later separated by collagen. A similar picture is seen in adult wound healing when new dermis is formed by fibrocytes which divide rapidly and move in from surrounding skin.

VERTEBRATE FOETAL EPIDERMIS

The epidermis is first a single cell layer but later becomes two or three cells in depth. The outermost cell layer, the periderm, even in mammals has a border of secretory microvilli covered by a thin cuticle-like material made up of fibrous and amorphous substances similar to the adult fish epidermal microvilli and cuticle. Later in mammalian development, the epidermal cells underneath become keratinised and periderm cells are shed individually into the amniotic fluid. Samples of these cells can be collected and cultured for detection of chromosomal abnormalities in the human foetus.

Sensory receptor cells in primitive animals are specialised epidermal cells, and probably the multicellular receptors of birds and mammals are similarly derived.

METAMORPHOSIS IN INSECTS

The epidermis in many insects is autolysed and replaced at metamorphosis from cells in the imaginal discs. Reconstruction therefore requires a return to embryonic organisation.

INDUCTIVE MECHANISMS (BELLAIRS 1971)

EMBRYONIC ORGANISERS

Embryonic induction mechanisms are sometimes reciprocal. Thus, in the chick embryo an apical ectodermal ridge is induced over wing bud mesoderm. Later, both the competence of the ectoderm to respond to the mesoderm and the ability of the mesoderm to induce this change are lost. The roles are now reversed, and the ectodermal ridge induces further development in the wing bud mesoderm. Without the covering ectoderm (future epidermis), further wing development ceases, and if a second ectodermal ridge is grafted next to the original ridge, two wings will grow from the original mesenchyme.

The complex interaction between ectoderm and mesoderm is also shown in the newt embryo when tail mesodermal organiser is grafted into the head region and there induces head structures. In tissue culture, induction occurs even when organiser and competent cells are separated by a filter which permits diffusion of substances but prevents cell contact. Probably some chemical inducing agent is liberated by the organiser and affects competent cells. However, practically anything from inorganic salts to dead tissue is capable of triggering off changes. Nevertheless, it is perhaps significant that there are subtle differences in the brain structure induced by other substances compared with the normal archenteral mesenchyme action. This suggests that, although induction can be produced by a variety of agents, genetically determined inducing agents are probably produced by organisers which trigger off changes in competent cells.

LATER MODULATORY MECHANISMS

Regional differences in the skin and formation of epidermal appendages are determined by inductive mechanisms between the dermis and a perpetually competent epidermis which start in the late foetus and continue throughout adult life. This differs from early embryonic induction in that the latter occurs only during a brief period of competence. In the three-day chick embryo (foetus), foot-forming mesoderm grafted under wing epidermis induces the latter to form scales and claws. The epidermis responds to wing mesoderm by formation of wing feathers and to thigh mesoderm by thigh feathers. In the adult chicken, rat and mouse, con-

tinuous induction of changes in the epidermis is exercised by the superficial dermis. Dermal papilla cells can be removed from an adult feather follicle and transplanted beneath the epidermis in another site. The epidermal follicle without its papilla atrophies, and in the new site a feather follicle is induced to form from the epidermis. The only exception to this rule is that feather follicles cannot be induced in normally apterous sites. This, Cohen (1969) suggests, is due to the opposing inductive effects of the superficial apterous dermis and the transplanted feather papilla cells.

Hair dermal papilla cells have been similarly transplanted in adult rats and mice and induce not only formation of an epidermal hair follicle from the epidermis, but also onset of hair growth.

McLoughlin (1961) showed in chick embryonic skin in tissue culture that epidermal keratinisation is dependent on the presence of fibrocytes, and those from a particular site direct modulation of the epidermis along a particular path.

Billingham and Silvers (1965) separated adult cavy epidermis from the dermis in excised skin by means of trypsin. They then grafted the separated epidermis from one site on to the separated dermis from another site to form a recombination homograft which was replaced in a site where the dermis had been removed. The epidermis then took on the characteristics induced by the dermis combined with it. In this way, a change to sole epidermis could be induced in the epidermis of the back by foot-pad dermis.

Skin grafts in plastic surgery always retain the characteristics of the donor site because even in thin grafts a layer of donor dermis is retained. Even dermis with killed cells induces changes in epidermis, probably due to agents secreted earlier.

In the foetal mouse, primordia of the various hair follicles appear at different times during development, which indicates that the dermal papillae become functional in a chronological sequence. In the antlers of deer, new hair follicles are formed in this way each spring. In regenerated lizard tails new epidermal scales are formed, but here the scale pattern is not the same as the original.

Whatever the nature of embryonic organisers, they switch on certain genes permanently in competent cells. Continuous modulation by the dermis involves various effects. Some of these, as in induction of hair follicle and enamel organ development, resemble the organiser, but others, such as initiation of hair growth, appear different.

In the adult integument, modulatory changes in epidermis are clearly reversible, although organisation as epidermis is irreversibly determined. Modulations are also produced by hormones, as in frog epidermis which keratinises at metamorphosis in response to thyroxine. Vitamin A induces

a mucous change in chick embryonic ectoderm, but once the epidermis has differentiated and become keratinised no such effect is produced. Thereafter the effect is on modulation of cornification, seen in the induction by vitamin A of a keratohyalin granular layer in adult mouse tail scales with associated changes in the horny layer (Jarrett and Spearman 1964).

THE NATURE OF SKIN PATTERNS

Genetically determined skin patterns, such as in the positions of thoracic bristles in the fruit fly *Drosophila* or in the arrangement of reptilian scales, are dependent on an underlying dermal pre-pattern, itself dependent on the sites taken up by particular neural crest and mesodermal cells which migrate in the embryo. We do not know how these cells are directed, except that migration occurs along paths of least resistance and stops when cells come up against other cells. This suggests a distribution pattern of cells before migration begins so that they are in a position to move along particular routes. The morphogenesis of a complex structure such as a mammalian hair follicle from epidermis takes place in an ordered sequence of interrelated events. Thus, in mutants deficient in medullary cells, developing cortical cells move to occupy the vacant sites. Development is easier to comprehend in the much simpler bristle organ of the fruit fly. Stern (1954) found that in bristle development a fixed number of giant cells appears in the epidermis from an original sheet of morphologically similar cells. The points at which the giant cells appear are genetically determined. The large cells divide first to give a neuroreceptor cell and an upper cell, which divides again to give two cells both of which secrete cuticular material. One of these cuticular cells normally secretes a bristle and the other forms a collar with which it articulates. In one mutant, absence of bristles is due to formation of two socket cells instead of a bristle cell and socket cell, and other mutants effect other stages in bristle development. Clearly, the morphogenic sequence and interrelationships in the formation of a much more complex structure, such as a feather, can easily be upset by the failure of only one stage of differentiation. This was originally shown by H. Grüneberg in mutants effecting organogenesis in the house mouse.

Morphological changes in denervated lizard skin suggest that epidermal patterns are influenced by the cutaneous nerves. Intact nerves are also necessary for regeneration of limbs in newts, although not for embryonic limb development. If the number of nerve axons to an amputated lizard limb is increased by surgical means, some regeneration occurs, although this does not normally happen in reptiles (Singer 1968). The exact action of cutaneous nerves on epidermis is not clear, but there is a suggestion that some growth factor for epidermis is secreted by nerve endings.

Another possible association is between melanocytes and epidermal cells with a particular type of keratinisation. An example is the pigmented papule on each scale of the crocodile which undergoes a different type of cornification to the scale and suggests an interaction between pigment cells and epidermis. Melanosomes are phagocytosed by epidermal cells and inducing agents could also be transferred to them from melanocytes.

Camouflage patterns such as the zebra's stripes and leopard's spots, warning patterns which deter other animals such as the yellow bands on a wasp's abdomen, and accessory sexual colouration such as the bright colours of many male birds, have been evolved through natural selection. A wealth of information is available on the genetical control of melanin formation in mammals and on the genes which modify or suppress hair colour (Searle 1968), but how colour patterns are reproduced with pigment formed in some sites and not in others, or only at a certain time such as during hair growth, is obscure.

The way concentric rings are laid down in tortoise carapace scales suggests that mitotically active zones occur around each scale boundary with falling gradients towards the centres. This could be determined by gradients in chalones, in cyclic AMP, or in adenyl cyclase activity. Gradients of mitotic activity are basic to all morphological patterns and are subject to dermal modulation.

INACTIVATION AND ACTIVATION OF GENES

Jacob and Monod in 1961, as a result of investigations on bacterial mutants, introduced the concept of coordinate gene repression to explain control of enzyme systems. A crude method is feedback inhibition, in which the accumulated product of enzyme action inhibits the enzyme from working. Finer control is by coordinate repression. In this mechanism, a group of structural genes for a sequence of enzymes in a metabolic reaction arranged along a DNA chain is controlled by a single operator region located at one end, the initiation point for RNA transcription. The operator plus its group of structural genes is termed an operon. In an unlinked region is a regulatory gene, whose product is a repressor molecule which reacts with the product of the structural genes and enables it to engage with the operator template and inactivate it. The mechanism is readily reversible, and a decrease in metabolites leads to release of the repressor so that transcription recommences. The operon model is attractively simple and possibly functions in all cells to control protein synthesis.

The nuclei of multicellular animals contain considerably more DNA than bacteria, and it is reasonable to suppose that the extra DNA is present in regulator genes required for differentiation. Indeed, multicellular animals produce relatively small amounts of RNA compared with their total DNA

even allowing for the fact that most structural genes are non-functional. Activation of genes occurs only in competent cells, and blocking of unwanted genes in differentiated cells (gene restriction) is permanent. Therefore, the binding of blocking molecules to DNA must be more stable than in coordinate repression. Certain nucleotides bind firmly to DNA, and here the intranuclear RNA is a possible source of blocking agents.

All the structural genes not directly concerned with general metabolism are probably restricted in this way in the egg, and in the embryo groups of these genes may be activated in turn by removal of blocking agents. On this hypothesis, competence is a brief period during which certain blocking agents can be removed by the action of an organiser on certain regulatory genes. It involves interaction of nucleus and cytoplasm.

HOMOLOGY

The implication in the concept of homologous structures is that they are similar because they are derived from a type of development which occurred in some common ancestor. This in turn implies that there must be a common genetic lineage from the ancestor. Yet it is exactly this genetic relationship which is denied by geneticists today. What is to be made of this paradox?

A complex structure, such as a hair follicle, owes its development not only to a very large number of genes but also to different types of genes. Those in different animals for keratin, tropocollagen, melanin, particular hormones and chalones are almost certainly homologous, although occasional mimic genes occur. Genes for production of inductive agents by particular organisers which seem only to trigger off events in competent cells need not be homologous, but those which determine the timing and duration of competence, one suspects, are homologous. On the other hand, there are numerous other genes for metabolic enzymes which indirectly determine rates of growth and cell movement, and therefore whether organisers and competent cells occur in the right places at the right time. These genes may have a profound effect on development, but in many instances the metabolic pathway can be taken over easily by an alternative route with different enzymes formed by different sets of genes. In comparative histochemical studies of the integument, different animals often have quite different enzymes as well as different isoenzymes. Examples are the various dehydrogenases and phosphatases which occur. Thus, human Langerhans cells contain ATPase, but alkaline phosphatase occurs in these cells in the potto. Selection out of a particular metabolic gene action will occur if it has some detrimental effect elsewhere in the body. It may then be replaced by another gene operon with a different set of enzymes, although the final morphogenetic change is not altered if

advantageous. Many genes in homologous structures must be of this type and therefore unrelated. Some structures, such as the trichocyst pattern in *Paramecium*, are inherited cytoplasmically, and similar effects probably occur in Metazoa. Clearly, therefore, most genes in anatomically homologous structures are not homologous, but it would be surprising if none were related, as has been suggested. Confusion has arisen because evolutionary morphologists think in terms of complete homology which never occurs in nature. One must think of all homologous organs as only partially homologous.

Some structures with little selective advantage, such as the preen gland in pigeons, occur only in a proportion of individuals in the species. This is due to interaction with the genetic background which varies from individual to individual and modifies gene expression.

CONCLUSIONS

In this chapter many questions have been posed, but only a few tentative answers suggested. The convergence of research in embryology, genetics, biophysics, biochemistry, immunology and cell biology provides new information almost every week on how cells develop. It is a subject which is likely to expand and introduce important new biological concepts within the next few years. The integument in both invertebrates and vertebrates has been found particularly suitable for such studies (Bellairs 1971; Billingham and Silvers 1965; De Beer 1971).

SUMMARY

Embryonic organisation occurs in sequences of stages. Epidermal development is determined in the embryo by contact with particular types of mesenchyme. The adult dermal cells continue to control the type of epidermal organisation and development of appendages throughout life, in contrast to the transient period of competence of embryonic cells to organisers.

19

GENERAL CONCLUSIONS

The integuments of invertebrates and vertebrates show similarities in function which exceed obvious differences. Both are continually replaced from dividing germinal cells, and common secretory products include melanin, collagen and various muco-substances. An exosecreted cuticle is found in many invertebrates and also in lower vertebrates, but remnants of the original cuticle appear to occur even in mammals, as in human foetal skin and the material between the keratinised cells of the adult stratum corneum. Indeed, the finding of a cuticle over the horny layer in Amphibia suggests that keratinocytes were derived from cuticularcytes. Glandular cells and receptor cells, which occur individually in lower animals, are mostly grouped together as multicellular glands and sensory end organs in higher vertebrates, and so the epidermis in these species is almost entirely composed of keratinocytes.

Sclerotisation is confined to invertebrates and keratinisation has only been found in vertebrates; both are processes which strengthen the proteinous covering of the skin. The periodic sloughing of this outer covering is necessary for growth of the body and in both invertebrates and vertebrates is often (perhaps always) effected hormonally, and is frequently dependent on impulses from the central nervous system to endocrine glands. Similar control mechanisms are involved in active colour changes in skin.

A thick cuticle is in some ways more efficient than the keratinised cellular layer of vertebrates. Thus, in arthropods and nematodes much of the protein material is reabsorbed prior to ecdysis, but in moult of keratinised structures none is conserved.

Cilia have been lost in vertebrate skin although they are retained in internal epithelia. A frequent feature of the integument is the formation of a mineralised skeleton, deposited as crystalline salts on organic micellae, in cuticles, in keratinised cells and in the dermis.

The skin, because it is superficial and readily observed, is useful for investigation of a wide range of physiological processes. One example is the skin colour changes in the octopus, useful in studies on the central nervous system. For the experimentalist, the integument clearly has scope far beyond the confines of Skin Biology.

CLASSIFICATION ADOPTED: LIST OF PHYLA

Protozoa
Porifera
Mesozoa
Cnidaria
Ctenophora
Platyhelminthes
Nemertina
Nematoda
Gastrotricha
Kinorhyncha
Nematomorpha
Rotifera
Acanthocephala
Priapulida
Entoprocta
Ectoprocta
Brachiopoda
Chaetognatha
Phoronidea
Echiurida
Sipunculoidea
Annelida
Pogonophora
Echinodermata
Arthropoda
Mollusca
Hemichordata
Tunicata
Chordata

BIBLIOGRAPHY

The following works contain additional references.

FURTHER READING

Bellairs, A. d'A. (1969). *The Life of Reptiles*, vol. 2. Weidenfeld and Nicolson: London.

Breathnach, A. S. (1971*b*). *Melanin Pigmentation of the Skin*. Oxford University Press: London.

Chapman, R. F. (1969). *The Insects: Structure and Function*. English Universities Press: London.

Fraser, R. D. B. (1969). Keratins. *Scient. Am.* **221**, 86 (Offprint No. 1155.)

Fraser, R. D. B., MacRae, T. P., and Rogers, G. E. (1972). *Keratins, Their Composition, Structure and Biosynthesis*. Thomas: Chicago.

Graham, A. (1957). The molluscan skin with special reference to prosobranchs. *Proc. malac. Soc.* **32**, 135.

Greenwood, P. H. (1963). '*J. R. Norman's 'A History of Fishes*'. Benn: London. 2nd ed.

Hadzi, J. (1963). *The Evolution of the Metazoa*. Pergamon Press: Oxford.

Jarrett, A., Spearman, R. I. C., and Riley, P. A. (1966). *Dermatology. A Functional Introduction*. English Universities Press: London.

Lee, D. L. (1966). The structure and composition of the helminth cuticle. *Adv. Parasitol.* **4**, 187.

Montagna, W., and Parakkal, P. F. (1973). *The Structure and Function of Skin*. Academic Press: New York and London. 3rd ed.

Munro Fox, H., and Vevers, G. (1960). *The Nature of Animal Colours*. Sidgwick and Jackson: London.

Wigglesworth, V. B. (1965). *Principles of Insect Physiology*. Methuen: London. 6th ed.

REFERENCES

Abolinŝ-Krogis, A. (1968). Shell regeneration in *Helix pomatia*, with special reference to the elementary calcifying particles. P. 75 in *Studies in the Structure, Physiology and Ecology of Molluscs*. Ed. V. Fretter (Symposium of the Zoological Society of London No. 22). Academic Press: New York and London.

Alexander, P., Hudson, R. F., and Earland, C. (1963). *Wool, its Chemistry and Physics*. Chapman and Hall: London. 2nd ed.

Allee, W. C., Park, O., Emerson, A. E., Park, T., and Schmidt, K. P. (1949). *Principles of Animal Ecology*. Saunders: Philadelphia.

Ambrose, E. J., and Easty, D. M. (1970). *Cell Biology*. Nelson: London.

Andersen, S. O., and Weis-Fogh, T. (1964). Resilin, a rubber-like protein in arthropod cuticle. *Adv. Insect Phys.* **2**, 1.

Atkins, D. (1932). The cilliary feeding mechanism of the entoproct polyzoa and a comparison with that of ectoproct polyzoa. *Quart. Jl microsc. Sci.* **75** (NS), 393.

Bagnara, J. T., and Hadley, M. E. (1969). The control of bright coloured pigment cells of fishes and amphibians. *Am. Zool.* **9**, 465.

Bancroft, J. D. (1967). *An Introduction to Histochemical Technique.* Butterworth: London.

Barnett, S. A. (1965). Adaptation of mice to cold. *Biol. Rev.* **40**, 5.

Barrington, E. J. W. (1963). *An Introduction to General and Comparative Endocrinology.* Oxford University Press: London.

Barrington, E. J. W. (1967). *Invertebrate Structure and Function.* Nelson: London.

Barrington, E. J. W., and Barron, N. (1960). On the organic binding of iodine in the tunic of *Ciona intestinalis. J. mar. biol. Ass. U.K.* **39**, 513.

Barrington, E. J. W., and Thorpe, A. (1968). Histochemical and biochemical aspects of iodine binding in the tissue of the ascidian *Dendrodora grossularia* (Van Beneden). *Proc. R. Soc. London. B* **171**, 91.

Beklemishev, W. N. (1969). *Principles of Comparative Anatomy of Invertebrates.* Oliver and Boyd: Edinburgh.

Beedham, G. E. (1958). Observations on the mantle of the Lamellibranchia. *Quart. Jl microsc. Sci.* **99**, 181.

Bellairs, A. d'A. (1969). *See* 'Further Reading'.

Bellairs, R. (1971). *Developmental Processes in Higher Vertebrates.* Logos Press: London.

Bentley, P. J., and Schmidt-Nielsen, K. (1966). Cutaneous water loss in reptiles. *Science, N.Y.* **151**, 1547.

Bereiter-Hahn, J. (1971). Licht und elektronmikroskopische Untersuchungen zur Funktion von Tonofilamenten in der Epidermiszellen von Fischen. *Cytobiologie* **4**(1), 73.

Biedermann, W. (1926). Vergleichende Physiologie des Integuments der Wirbeltiere. *Ergebn. Biol.* **1**, 1.

Billingham, R. E., Mangold, R., and Silvers, W. K. (1959). Neogenesis of hair follicles in adult skin; the neogenesis of skin in the antlers of deer. *Ann. N.Y. Acad. Sci.* **83**, 491.

Billingham, R. E., and Medawar, P. B. (1951). The technique of free skin grafting in mammals. *J. exp. Biol.* **28**, 385.

Billingham, R. E., and Silvers, W. K. (1965). Some unsolved problems in the biology of skin. Ch. 1 in *Biology of the Skin and Hair Growth.* Eds. A. G. Lyne, and B. F. Short. Angus and Robertson: Sydney.

Bird, A. F. (1971). *The Structure of Nematodes.* Academic Press: New York and London.

Bissonnette, T. H., and Wilson, E. (1939). Shortening daylight periods between May 15 and September 12 and the pelt cycle of the mink. *Science, N.Y.*, **89**, 418.

Bligh, J. (1966). The thermosensitivity of the hypothalamus and thermoregulation in mammals. *Biol. Rev.* **41**, 317.

Breathnach, A. S. (1971*a*). *An Atlas of The Ultrastructure of Human Skin.* Churchill: London.

Breathnach, A. S. (1971*b*). *See* 'Further Reading'.

Brodal, A., and Fange, R. (1963). *The Biology of Myxine.* Scandinavian University Books: Oslo.

Brody, I. (1962). The ultrastructure of the horny layer in normal and psoriatic epidermis as revealed by electron microscopy. *J. invest. Derm.* **39**, 519.

Brody, I. (1966). Intercellular space in normal human stratum corneum. *Nature, Lond.* **209**, 472.

Brown, C. H. (1950). Quinone tanning in the animal kingdom. *Nature, London* **165**, 275.

Brunet, P. C. J., and Carlisle, D. B. (1958). Chitin in Pogonophora. *Nature, Lond.* **182**, 1689.

Bullough, W. S. (1962). The control of mitotic activity in adult mammalian tissues. *Biol. Rev.* **37**, 307.

Bullough, W. S., and Laurence, E. (1964). Mitotic control by internal secretion: The role of the chalone adrenalin complex. *Exp. Cell Res.* **33**, 176.

Carlisle, D. B., and Knowles, F. (1959). *Endocrine Control in Crustaceans.* Cambridge University Press: London.

Carthy, J. D., and Newell, J. E. (Eds.) (1968). *Invertebrate Receptors* (Symposium of the Zoological Society of London No. 23). Academic Press: New York and London.

Causey-Whittow, G. (Ed.) (1970). *Comparative Physiology of Thermoregulation.* Academic Press: New York and London.

Chapman, R. F. (1969). *See* 'Further Reading'.

Chase, H. B., and Silvers, A. F. (1969). The biology of hair growth. Ch. 1 in *The Biological Basis of Medicine*, vol. 6. Eds. E. D. Bittar and N. Bittar. Academic Press: New York and London.

Chavin, W. (1969). Fundamental aspects of morphological melanin colour changes in vertebrate skin. *Am. Zool.* **9**, 505.

Coggeshall, R. E. (1966). A fine structural analysis of the epidermis of the earthworm *Lumbricus terrestris* L. *J. Cell Biol.* **28**, 95.

Cohen, J. (1969). Interactions in the skin. *Br. J. Derm.* **81**, Suppl. 3, 46.

Dales, P. (1967). *Annelids.* Hutchinson: London.

Davson, H. (1970). *A Text Book of General Physiology.* Churchill: London.

Dawson, A. B. (1920). The integument of *Necturus maculosus. J. Morph.* **34**, 487.

De Beer, G. R. (1971). *Homology, an Unsolved Problem.* Oxford University Press: London.

Dennell, R. (1960). Integument and exoskeleton. Ch. 14 in *The Physiology of Crustacea*, vol. 1. Ed. T. H. Waterman. Academic Press: New York and London.

De Robertis, E. D. P., Nowinski, W. W., and Saez, F. A. (1965). *Cell Biology.* Saunders: Philadelphia. 4th ed.

Dijkgraaf, S. (1963). The functioning and significance of the lateral-line organs. *Biol. Rev.* **38**, 51.

Dilly, P. N. (1969). The ultrastructure of the test of the tadpole larva of *Ciona intestinalis. Z. Zellforsch. mikrosk. Anat.* **95**, 331.

Donovan, B. T., and Harris, G. W. (1954). Effect of pituitary stalk section on light-induced oestrus in the ferret. *Nature, Lond.* **174**, 503.

Doyle, W. L., and Gorecki, D. (1961). The so-called chloride cell of the fish gill. *Physiol. Zool.* **34**, 81.

Ebeling, W. (1964). The permeability of insect cuticle. Ch. 9 in *The Physiology of Insecta*, vol. 3. Ed. M. Rockstein. Academic Press: New York and London.

Ebling, F. J. (1965). Systemic factors affecting the periodicity of hair follicles. Ch. 31 in *Biology of the Skin and Hair Growth.* Eds. A. G. Lyne and B. F. Short. Angus and Robertson: Sydney.

Ebling, F. J., and Johnson, E. (1961). Systemic influence on activity of hair follicles in skin homografts. *J. Embryol. exp. Morph.* **9**, 285.

Elden, H. R. (Ed.) (1971). *Biophysical Properties of the Skin.* Wiley: New York.

Ellis, V. L., Ross, D. M., and Sutton, L. (1969). The pedal disc of the swimming sea anemone *Stomphia coccinea* during detachment, swimming and resettlement. *Can. J. Zool.* **47**, 333.

Farbman, A. (1966). Plasma membrane changes during keratinization. *Anat. Rec.* **156**, 269.

Farquhar, M. G., and Palade, G. E. (1965). Cell junctions in amphibian skin. *J. Cell Biol.* **26**, 263.

Filshie, B. K., and Rogers, G. E. (1962). An electron microscope study of the fine structure of feather keratin. *J. Cell Biol.* **13**, 1.

Fingerman, M. (1969). Cellular aspects of the control of physiological colour change in crustaceans. *Am. Zool.* **9**, 443.

Fitzpatrick, T. B., Masamitsu, M., and Ishikawa, K. (1967). The evolution of concepts of melanin biology. Ch. 1 in *Advances in the Biology of Skin.* 8. *The Pigmentary System.* Eds. W. Montagna and F. Hu. Pergamon Press: Oxford.

Florey, E. (1969). Ultrastructure and function of cephalopod chromatophores. *Am. Zool.* **9**, 429.

Florkin, M., and Scheer, B. T. (Eds.) (1968). *Chemical Zoology.* Vol. 2. *Porifera, Coelenterata and Platyhelminthes.* Academic Press: New York and London.

Florkin, M., and Scheer, B. T. (Eds.) (1969*a*). *Chemical Zoology.* Vol. 3. *Echinodermata, Nematoda and Acanthocephala.* Academic Press: New York and London.

Florkin, M., and Scheer, B. T. (Eds.) (1969*b*). *Chemical Zoology.* Vol. 4. *Annelida, Echiura and Sipuncula.* Academic Press: New York and London.

Fogal, W., and Fraenkel, G. (1970). Histogenesis of the cuticle of the adult flies *Sarcophaga bullata and S. argyrostoma. J. Morph.* **130**, 140.

Frankfurt, O. S. (1971). Epidermal chalone effects on cell cycle and on development of hyperplasia. *Exp. Cell Res.* **64**, 136.

Fraser, R. D. B. (1969). *See* 'Further Reading'.

Fraser, R. D. B., MacRae, T. P., and Rogers, G. E. (1972). *See* 'Further Reading'.

Fretter, V. (1952). Experiments with P 32 and I 131 on species of *Helix*, *Arion* and *Agriolimax. Quart. Jl microsc. Sci.* **93**, 133.

Fretter, V., and Graham, A. (1962). *British Prosobranch Molluscs.* Ray Soc.: London.

Friedrich, V. L., and Langer, R. M. (1969). Fine structure of cribellate spider silk. *Am. Zool.* **9**, 91.

Fujii, R., and Novales, R. R. (1969). Cellular aspects of the control of physiological colour change in fishes. *Am. Zool.* **9**, 453.

Fukuyama, K., and Epstein, W. L. (1969). Sulfur-containing proteins and epidermal keratinization. *J. Cell Biol.* **40**, 830.

Gilchrist, J. D. F. (1920). Ecdysis in a teleostean fish, *Agriopus. Q. Jl microsc. Sci.* **64**, 575.

Gillespie, J. M. (1965). The high sulphur proteins of normal and aberrant keratins. Ch. 23 in *Biology of the Skin and Hair Growth.* Eds. A. G. Lyne and B. F. Short. Angus and Robertson: Sydney.

Gillespie, J. M. (1970). Mammoth hair: Stability of alpha keratin, structure and constituent proteins. *Science, N.Y.* **170**, 1100.

Goreau, T. F. (1963). Calcium carbonate deposition by coraline algae and corals in relation to their roles as reef builders. *Ann. N.Y. Acad. Sci.* **109**, 1, 127.

Graham, A. (1957). *See* 'Further Reading'.

Greenwood, P. H. (1963). *See* 'Further Reading'.

Grimstone, A. V. (1961). Fine structure and morphogenesis in Protozoa. *Biol. Rev.* **36**, 97.

Gross, J. (1961). Collagen. *Scient. Am.* **204**, 120. (Offprint No. 88.)

Hadley, M. E., and Goldman, J. M. (1969). Physiological colour changes in reptiles. *Am. Zool.* **9**, 489.

Hadzi, J. (1963). *See* 'Further Reading'.

Hare, P. E. (1963). Amino acids in the proteins from calcite and aragonite in the shells of *Mytilus californianus. Science, N.Y.* **139**, 216.

Harkness, R. D. (1961). Biological functions of collagen. *Biol. Rev.* **36**, 399.
Harkness, R. D. (1968). Mechanical properties of collagenous tissues. Ch. 6 in *Treatise on Collagen*, vol. 2A. Ed. B. S. Gould. Academic Press: New York and London.
Holbrow, E. J. (1970). *An ABC of Modern Immunology.* Lancet Publications: London.
Hughes, T. E. (1959). The cuticle of *Acarus siro* L. *J. exp. Biol.* **36**, 363.
Hunt, S. (1970). *Polysaccharide Protein Complexes in Invertebrates.* Head Press: London.
Huxley, J. S. (1963). *Evolution, the Modern Synthesis.* Allen and Unwin: London. 2nd ed.
Hyman, L. H. (1940). *The Invertebrates.* Vol. 1. *Protozoa through Ctenophora,* McGraw-Hill: New York.
Hyman, L. H. (1951*a*). *The Invertebrates.* Vol. 2. *Platyhelminthes and Rhynchocoela.* McGraw-Hill: New York.
Hyman, L. H. (1951*b*). *The Invertebrates.* Vol. 3. *Acanthocephala, Aschelminthes and Entoprocta.* McGraw-Hill: New York.
Hyman, L. H. (1955). *The Invertebrates.* Vol. 4. *Echinodermata, the Coelomate Bilateria.* McGraw-Hill: New York.
Hyman, L. H. (1958). Occurrence of chitin in the lophophorate phyla. *Biol. Bull. Woods Hole*, **114**, 106.
Hyman, L. H. (1959). *The Invertebrates.* Vol. 5. *Smaller Coelomate Groups.* McGraw-Hill: New York.
Hyman, L. H. (1967). *The Invertebrates.* Vol. 6. *Mollusca* 1. McGraw-Hill: New York.
Iggo, A. (1968). Electrophysiological and histological studies of cutaneous mechanoreceptors. Ch. 5 in *The Skin Senses.* Ed. D. R. Kenshalo. Thomas: Chicago.
Ivanov, A. V. (1963). *Pogonophora.* Academic Press: New York and London.
Iversen, O. H. (1969). Chalones of the skin. P. 29 in *Homeostatic Regulators.* Eds. G. E. W. Wolstenholme and J. Knight. Churchill: London.
Jaccarini, V., Bannister, W. H., and Micallef, H. (1968). The pallial glands and rock boring in *Lithophaga lithophaga. J. Zool., Lond.* **154**, 397.
Jakubowski, M. (1960). The structure and vascularisation of the skin of the eel *Anguilla anguilla* and the viviparous blenny *Zoarces viviparus. Acta biol. crac.* **3**, 1.
Jarrett, A. (Ed.) (1973). *Physiology and Pathophysiology of the Skin,* vol. 1. Academic Press: New York and London.
Jarrett, A., and Hardy, J. A. (1957). The value of alcohol for fixation of skin. *Stain Technol.* **32**, 225.
Jarrett, A., and Spearman, R. I. C. (1964). *Histochemistry of the Skin: Psoriasis. A Monograph on Normal and Abnormal Parakeratotic Epidermal Keratinization.* English Universities Press: London.
Jarrett, A., and Spearman, R. I. C. (1966). The histochemistry of the human nail. *Arch. Derm., Chicago* **94**, 652.
Jarrett, A., Spearman, R. I. C., and Riley, P. A. (1966). *See* 'Further Reading'.
Kann, S. (1926). Die Histologie der Fischhaut von biologischen Gesichtspunkten betrachtet. *Z. Zellforsch. mikrosk. Anat.* **4**, 482.
Kay, D. H. (1965). *Techniques for Electron Microscopy.* Blackwell: Oxford. 2nd ed.
Kelly, D. E. (1966). The Leydig cell in larval amphibian epidermis. *Anat. Rec.* **154**, 685.
Kerr, T. (1952). The scales of primitive living actinopterygians. *Proc. zool. Soc. Lond.* **122**, 55.
Kerr, T. (1955). The scales of modern lungfish. *Proc. zool. Soc. Lond.* **125**, 335.

Kitzan, S. M., and Sweeny, P. R. (1968). A light and electron microscope study of the structure of *Protopterus annectens* epidermis 1. Mucus production. *Can. J. Zool.* **46**, 767.

Koehler, J. K. (1965). A fine-structure study of the rotifer integument. *J. Ultrastruct. Res.* **12**, 113.

Kudo, R. R. (1966). *Protozoology*. Thomas: Illinois. 5th ed.

Landsborough Thomson, A. (1964). *A New Dictionary of Birds*. Nelson: London.

Larsen, O. (1973). Physiology of moulting. In *Physiology of Amphibia*, vol. 2. Ed. B. Lofts. Academic Press: New York and London. 2nd ed.

Lee, D. L. (1966). *See* 'Further Reading'.

Lee, D. L. (1970). The ultrastructure of the cuticle of adult female *Mermis nigrescens* (Nematoda). *J. Zool., Lond.* **161**, 513.

Lentz, T. L. (1966). *The Cell Biology of Hydra*. North Holland: Amsterdam.

L'Hélias, C. L. (1970). Chemical aspects of growth and development in insects. Ch. 12 in *Chemical Zoology*, vol. 5. *Arthropoda* A. Eds. M. Florkin and B. T. Scheer. Academic Press: New York and London.

Lillie, F. R. (1942). On the development of feathers. *Biol. Rev.* **17**, 247.

Lillie, R. D. (1966). *Histopathologic, Technic and Practical Histochemistry*. McGraw-Hill: New York.

Ling, J. K. (1970). Pelage and moulting in wild mammals with special reference to aquatic forms. *Quart. Rev. Biol.* **45**, 16.

Locke, M. (1965). The structure of septate desmosomes. *J. Cell Biol.* **25**, 166.

Lockwood, A. P. M. (1968). *Aspects of the Physiology of Crustacea*. Oliver and Boyd: Edinburgh.

Lotmar, W., and Picken, L. E. R. (1950). A new crystallographic modification of chitin and its distribution. *Experientia* **6**, 58.

McLoughlin, C. B. (1961). The importance of mesenchymal factors in the differentiation of chick epidermis. Modification of epidermal differentiation by contact with different types of mesenchyme. *J. Embryol. exp. Morph.* **9**, 385.

Maderson, P. F. A. (1965). Histological changes in the epidermis of snakes during the sloughing cycle. *J. Zool., Lond.* **146**, 98.

Malkinson, F. D. (1964). Permeability of the stratum corneum. Ch. 21 in *The Epidermis*. Eds. W. Montagna and W. C. Lobitz. Academic Press: New York and London.

Maloiy, G. M. (Ed.) (1972). *Comparative Physiology of Desert Animals* (Symposium of the Zoological Society of London No. 31). Academic Press: New York and London.

Mann, K. H. (1962). *Leeches (Hirudinea): Their Structure, Physiology and Embryology*. Pergamon Press: Oxford.

Mathews, M. B. (1967). Macromolecular evolution of connective tissue. *Biol. Rev.* **42**, 499.

Matoltsy, A. G., and Matoltsy, M. N. (1970). The chemical nature of keratohyalin granules of the epidermis. *J. Cell Biol.* **47**, 593.

Medawar, P. B. (1941). Sheets of pure epidermal epithelium from human skin. *Nature, Lond.* **148**, 783.

Meek, G. A. (1970). *Practical Electronmicroscopy for Biologists*. Wiley: London.

Miles, A. E. W. (Ed.) (1967). *The Structure and Chemical Organisation of Teeth*. 2 Vols. Academic Press: New York and London.

Miller, M. R., and Kasahara, M. (1967). Studies on the cutaneous innervation of lizards. *Proc. Calif. Acad. Sci.* **34**, 549.

Montagna, W. (Ed.) (1962). *Advances in the Biology of Skin*. Vol. 3. *Eccrine Sweat Glands and Eccrine Sweating*. Pergamon Press: Oxford.

Montagna, W., and Parakkal, P.F. (1973). *See* 'Further Reading'.

Montagna, W., and Ellis, R. A. (Eds.) (1961). *Advances in the Biology of Skin. Blood Vessels and Circulation.* Pergamon Press: Oxford.

Montagna, W., Ellis, R. A., and Silvers, A. F. (Eds.) (1963). *Advances in the Biology of Skin.* Vol. 4. *The Sebaceous Glands.* Pergamon Press: Oxford.

Montagna, W., and Hu, F. (Eds.) (1967). *Advances in the Biology of Skin.* Vol. 8. *The Pigmentary System.* Pergamon Press: Oxford.

Moritz, K., and Storch, V. (1970). Über den Aufbau des Integumentes der Priapuliden und der Sipunculiden. *Z. Zellforsch. mikrosk. Anat.* **105**, 55.

Munro Fox, H., and Vevers, G. (1960). *See* 'Further Reading'.

Niebauer, G. (1968). *Dendritic Cells of Human Skin.* Karger: Basel.

Noble, G. K. (1931). *The Biology of Amphibia.* Dover: New York.

Noble-Nesbitt, J. (1963). The fully formed intermoult cuticle and associated structures of *Podura aquatica* collembola. *Quart. Jl microsc. Sci.* **104**, 253.

Nordenskiöld, E. (1946). *The History of Biology.* Knopf: New York.

Olsson, R. (1961). The skin of *Amphioxus. Z. Zellforsch. mikrosk. Anat.* **54**, 90.

Oosten, J. van (1957). The skin and scales. P. 207 in *Physiology of Fishes*, vol. 1. Ed. M. E. Brown. Academic Press: New York and London.

Orrhage, L. (1971). Light and electron microscope studies of some annelid setae. *Acta. Zool.* **52**, 157.

Packard, A., and Sanders, G. (1969). What the octopus shows to the world. *Endeavour* **28**, 92.

Parakkal, P. F., and Matoltsy, A. G. (1964). A study of the fine structure of the epidermis of *Rana pipiens. J. Cell Biol.* **20**, 85.

Pautard, F. G. E. (1963). Mineralisation of keratin and its comparison with the enamel matrix. *Nature, Lond.* **199**, 531.

Peakall, D. B. (1969). Synthesis of silk mechanism and location. *Am. Zool.* **9**, 71.

Pearse, A. G. E. (1968). *Histochemistry: Theoretical and Applied.* Churchill: London. 3rd ed.

Pfeiffer, W. (1968). Über die Epidermis von *Latimeria chalumnae. Z. Morph. Ökol. Tiere* **63**, 419.

Pikkarainen, J., and Kulonen, E. (1969). Comparative chemistry of collagen. *Nature, Lond.* **223**, 839.

Pilkington, J. B. (1969). The organisation of skeletal tissues in the spines of *Echinus esculentus. J. mar. biol. Ass. U.K.* **49**, 857.

Pinkus, H. (1952). Examination of the epidermis by the strip method. *J. invest. Derm.* **19**, 431.

Pitelka, D. R. (1963). *Electron Microscopic Structure of Protozoa.* Pergamon Press: Oxford.

Popper, K. R. (1963). *Conjectures and Refutations.* Routledge and Kegan Paul: London.

Potts, W. T. W., and Parry, G. (1964). *Osmotic and Ionic Regulation in Animals.* Pergamon Press: Oxford.

Prosser, C. L., and Brown, F. A. (1961). *Comparative Animal Physiology.* Saunders: Philadelphia.

Pryor, M. G. M. (1962). Sclerotization. Ch. 8 in *Comparative Biochemistry*, vol. 4B. Eds. M. Florkin and H. S. Mason. Academic Press: New York and London.

Ralph, C. L. (1969). The control of colour in birds. *Am. Zool.* **9**, 521.

Ramachandran, G. N., and Gould, B, S. (Eds.) (1967). *Treatise on Collagen*, 3 vols. Academic Press: New York and London.

Renold, A. E., and Cahill, G. F. (1965). *Handbook of Physiology.* Vol. 5. *Adipose Tissue.* American Physiological Society: Washington, DC.

Richards, A. G. (1951). *The Integument of Arthropods*. University of Minnesota Press: Minneapolis.

Riley, P. A. (1967). A study of the distribution of epidermal dendritic cells in pigmented and unpigmented skin. *J. invest. Derm.* **48**, 28.

Riley, V. (Ed.) (1972). *Pigmentation. Its Genesis and Biologic Control.* Appleton-Century-Crofts: New York.

Riviera, J. A. (1962). *Cilia, Ciliated Epithelium and Ciliary Activity*. Pergamon Press: Oxford.

Robson, F. A. (1964). The cuticle of *Peripatopsis mosleyi. Quart. Jl microsc. Sci.* **105**, 281.

Rockstein, M. (Ed.) (1964). *The Physiology of Insecta*, 3 vols. Academic Press: New York and London.

Rogers, A. W. (1969). *Techniques of Autoradiography*. Elsevier: Amsterdam. 2nd ed.

Roitt, I. M., Greaves, M. F., Torrigiani, G., Brostoff, J., and Playfair, J. H. L. (1969). The cellular basis of immunological responses. *Lancet* **2**, 367.

Ross, R. (1968). The connective tissue fibre forming cell. Ch. 1 in *Treatise on Collagen*, vol. 2A. Ed. B. S. Gould. Academic Press: New York and London.

Rowden, G. (1968). Ultrastructural studies on keratinized epithelia of the mouse. *J. invest. Derm.* **51**, 51.

Rudall, K. M. (1947). X-ray studies on the distribution of protein chain types in the vertebrate epidermis. *Biochim. biophys. Acta* **1**, 549.

Rudall, K. M. (1963). The chitin protein complexes of insect cuticles. *Adv. Insect Phys.* **1**, 257.

Rudall, K. M. (1965). Skeletal structure in insects. P. 83 in *Aspects of Insect Physiology* (Biochem. Soc. Symp. No. 25). Ed. T. W. Goodwin. Academic Press: New York and London.

Ryder, M. L., and Stephenson, S. K. (1968). *Wool Growth.* Academic Press: New York and London.

Saleuddin, A. S. M., and Wilbur, K. M. (1969). Shell regeneration in *Helix pomatia. Can. J. Zool.* **47**, 51.

Scholander, P. F., Hock, R., Walters, V., and Irving, L. (1950). Adaptation to cold in arctic and tropical mammals and birds, in relation to body temperature, insulation and basal metabolism. *Biol. Bull., Woods Hole* **99**, 259.

Schultze, H. E., and Heremans, J. F. (1966). *Molecular Biology of Human Proteins*, vol. 1. Elsevier: Amsterdam.

Searle, A. G. (1968). *Comparative Genetics of Coat Colour in Mammals.* Logos Press: London.

Sinclair, D. (1967). *Cutaneous Sensation.* Oxford University Press: London.

Singer, M. (1968). Some quantitative aspects concerning the trophic role of the nerve cell. P. 233 in *Systems Theory Biology.* Ed. M. D. Mesarovic. Springer-Verlag: Berlin.

Slautterback, D. B., and Fawcett, D. W. (1959). The development of the cnidoblasts of *Hydra.* An electron microscope study of cell differentiation. *J. biophys. biochem. Cytol.* **5**, 441.

Sleigh, M. A. (1962). *The Biology of Cilia and Flagella.* Pergamon Press: Oxford.

Smith, D. S. (1968). *Insect Cells, Their Structure and Function.* Oliver and Boyd: Edinburgh.

Smyth, J. D. (1969). *Physiology of Cestodes.* Oliver and Boyd: Edinburgh.

Souza Santos, H. de, and Silva Sasso, W. da (1970). Ultrastructural and histochemical studies on the epithelium revestment layer in the tube feet of the starfish *Asterina stellifera. J. Morph.* **130**, 287.

Spearman, R. I. C. (1964). The evolution of mammalian keratinized structures.

P. 67 in *The Mammalian Epidermis and its Derivatives* (Symposium of the Zoological Society of London No. 12). Ed. F. J. Ebling. Academic Press: New York and London.

Spearman, R. I. C. (1966). The keratinization of epidermal scales, feathers and hairs. *Biol. Rev.* **41**, 59.

Spearman, R. I. C. (1968*a*). A histochemical examination of the epidermis of the southern elephant seal (*Mirounga leonina* L.) during the telogen stage of hair growth. *Aust. J. Zool.* **16**, 17.

Spearman, R. I. C. (1968*b*). Epidermal keratinization in the salamander and a comparison with other amphibia. *J. Morph.* **125**, 129.

Spearman, R. I. C. (1969). The epidermis of the gopher tortoise *Testudo polyphemus* (Daudin). *Acta Zool.* **50**, 1.

Spearman, R. I. C. (1970*a*). The epidermis and its keratinisation in the African elephant (*Loxodonta africana*). *Zool. Afrikana* **5**, 327.

Spearman, R. I. C. (1970*b*). Some light microscopical observations on the stratum corneum of the guinea pig, man, and the common seal. *Br. Jl Derm.* **83**, 582.

Spearman, R. I. C. (1971). The integumentary system. Ch. 25 in *Physiology and Biochemistry of the Domestic Fowl*, vol. 2. Eds. D. J. Bell and B. M. Freeman. Academic Press: New York and London.

Spearman, R. I. C., and Garretts, M. (1966). The site of deposition and rate of clearance of saline after subcutaneous injection into the mouse tail. *J. invest. Derm.* **46**, 251.

Spearman, R. I. C., and Riley, P. A. (1969). A comparison of the epidermis and pigment cells of the crocodile with those in two lizard species. *Zool. J. Linn. Soc. Lond.* **48**, 453.

Stern, C. (1954). Two or Three Bristles. *Am. Scient.* **42**, 213.

Steven, D. M. (1963). The dermal light sense. *Biol. Rev.* **38**, 204.

Storch, V., and Welsch, U. (1970). Über die Feinstruktur der Polychaeten Epidermis Annelida. *Z. Morph. Ökol. Tiere.* **66**, 310.

Sturkie, P. D. (1965). *Avian Physiology*. Cornell University Press: Ithaca. 2nd ed.

Tarlo, L. B. (1964). The origin of bone. Ch. 1 in *Bone and Tooth*. Ed. H. J. Blackwood. Pergamon Press: Oxford.

Thompson, D'Arcy W. (1942). *On Growth and Form*. Cambridge University Press: London. 2nd ed.

Thompson, S. W., and Hunt, R. D. (1966). *Selected Histochemical and Histopathological Methods*. Thomas: Chicago.

Timmermans, L. P. M. (1969). Studies on shell formation in molluscs. *Arch. néerl. Zool.* **19**, 417.

Travis, D. F. (1955). The moulting cycle of the spiny lobster *Panulirus argus latreille*. Ecdysal, histological and histochemical changes in the hepatopancreas and integumental tissues. *Biol. Bull., Woods Hole* **108**, 88.

Tregear, R. T. (1966). *Physical Functions of Skin*. Academic Press: New York and London.

Van Abbé, N. J., Spearman, R. I. C., and Jarrett, A. (1969). *Pharmaceutical and Cosmetic Products for Topical Administration*. Heinemann: London.

Voitkevich, A. A. (1966). *The Feathers and Plumage of Birds*. Sidgwick and Jackson: London.

Waring, H. (1963). *Colour Change Mechanisms of Cold Blooded Vertebrates*. Academic Press: New York and London.

Watson, G. E. (1963). Feather replacement in birds. *Science, N.Y.* **139**, 50.

Watson, M. R. (1958). The chemical composition of earthworm cuticle. *Biochem. J.* **68**, 416.

Weddell, G., Palmer, E., and Pallie, W. (1955). Nerve endings in mammalian skin. *Biol. Rev.* **30**, 159.

Weiner, J. S., and Hellmann, K. (1960). The sweat glands. *Biol. Rev.* **35**, 141.

Welsch, U. (1968). Beobachtungen über die Feinstruktur der Haut und des aüsseren Atrialepithels von *Branchiostoma lanceolatum* Pall. *Z. Zellforsch. mikrosk. Anat.* **88**, 565.

Whitear, M. (1970). The skin surface of bony fishes. *J. Zool. Lond.* **160**, 437.

Whitear, M. (1971*a*). The free nerve endings in fish epidermis. *J. Zool. Lond.* **163**, 231.

Whitear, M. (1971*b*). Cell specialisation and sensory function in fish epidermis. *J. Zool. Lond.* **163**, 237.

Wigglesworth, V. B. (1964). The hormonal control of growth and reproduction in insects. *Adv. Insect Phys.* **2**, 244.

Wigglesworth, V. B. (1965). *See* 'Further Reading'.

Wigglesworth, V. B. (1970). Structural lipids in the cuticle and the function of the oenocytes. *Tissue and Cell* **2**, 155.

Wilbur, K. M. (1964). Shell formation and regeneration. P. 243 in *Physiology of Mollusca*, vol. 1. Eds. K. M. Wilbur and C. M. Yonge. Academic Press: New York and London.

Wildman, A. B. (1954). *The Microscopy of Animal Textile Fibres.* Wool Ind. Res. Ann: Leeds.

Wiley, M. L., and Collette, B. B. (1970). Breeding tubercles and contact organs in fishes, their occurrence, structure and significance. *Bull. Am. Mus. nat. Hist.* **143**, 147.

Williams, A. (1956). The calcareous shell of the Brachiopoda and its importance to their classification. *Biol. Rev.* **31**, 243.

Zelickson, A. S. (1967). *Ultrastructure of Normal and Abnormal Skin.* Kimpton: London.

GLOSSARY

Anagen Onset of hair or feather growth.

Apocrine Secretion formed by budding off of cytoplasm from secretory cells.

Basal lamina (*basal membrane*) Not a true membrane but a layer of dermis (0.1–0.2 μm thick) immediately beneath the epidermis. It is rich in mucopolysaccharides and contains a network of fine collagen fibres. It is absent only in some primitive acoelomates.

Catagen Shrinkage stage of hair follicle following anagen.

Chalones Inhibitory agents to mitosis, exosecreted by tissue cells.

Chitin A fibrous neutral mucopolysaccharide containing glucosamine and acetyl glucosamine found in certain invertebrate cuticles.

Chromatophore A pigment-containing cell responsible for skin colouration.

Collagen A fibrous protein containing hydroxyproline, found in connective tissue and certain invertebrate cuticles.

Cuticle In invertebrates and lower vertebrates, this is formed as a hard exosecretion of the epidermis. It varies in composition in different groups.

Cuticularcytes Epithelial cells which exosecrete a cuticle.

Dermis (*corium or cutis*) The connective tissue layer beneath the epidermis.

Desmosome (*macula adhaerens*) Junctions in epidermis. By light microscopy seen as prickles between neighbouring epidermal cells. By electron microscopy they appear as disc-shaped attachments with the plasma membranes separated by a 240-Å space. Within the space is a plaque of dense material.

Ecdysis Shedding of the epidermal cuticle.

Elastin A fibrous protein of vertebrate connective tissue which shows elastic and tensile properties.

Endosecretion Retention of secreted substances in the cytoplasm. Protein-retaining cells have a smooth endoplasmic reticulum and free ribosomes.

Epicuticle The outer region of the insect cuticle which contains hydrophobic waxes. Also the non-keratinous layer over the mammalian hair cells.

Epidermal intercellular pathway The continuous pathway in between the epidermal cells through which water and dissolved substances diffuse. It connects with the dermal intercellular space. The pathway may be occluded by junctions in the superficial cells.

Fibroins Glycine-rich fibrous proteins found in arthropod cuticle and in silkmoth silk.

Ground substance The amorphous material rich in acid mucopolysaccharides precipitated between the collagen and elastin fibres in fixed tissue sections. In life, collagen and ground substance are associated as a fibrous gel.

Holocrine Secretion formed from whole cells.

Hyperkeratosis Formation over a granular layer of a thick stratum corneum not containing basophilic nuclear remnants. Seen in the elephant.

Hypodermis The loose subcutaneous connective tissue beneath the dermis in which fat is stored.

Keratin A complex high molecular weight protein containing both fibrous and globular components laid down in epidermal cells of vertebrates.

Keratinisation This comprises keratin synthesis and the breakdown of normal cell structure by autolysis. Epidermal keratinisation is sometimes termed cornification. Strictly this should be 'keratisation' from the Greek *keratos* for horn, but 'keratinisation' is now widely used.

Keratohyalin Complexes of protein, phospholipid and calcium which form basophilic cytoplasmic granules in the mammalian granular layer (stratum granulosum). Not found in other groups.

Langerhans cell Unpigmented dendritic cell found in the spaces between vertebrate epidermal cells and of unknown function and derivation.

Melanin A complex polymer of indole-5,6-quinone linked to a protein. A brown to black pigment. In vertebrates it is laid down inside melanocytes, but is formed outside cells in insect cuticles.

Melanocyte A melanin-forming cell.

Melanophore A melanocyte which contains melanosomes and functions as a chromatophore.

Melanosome The cytoplasmic organelle in which melanin is deposited.

Mineralisation The deposit of crystalline mineral salts in epidermal cuticles, shells and keratinised cells, and in dermal scales, bones and teeth. In this process, organic particles act as nuclei for crystallisation.

Muco-substances A variety of protein–carbonhydrate complexes which include mucopolysaccharides and glycoproteins.

Mucus A viscous or watery solution containing muco-substances, exosecreted by certain epithelial cells.

Parakeratosis Formation of a stratum corneum containing basophilic remnants of nuclei. Also found in pathological conditions.

Receptors Sensory cells in close contact with nerve endings and which convey stimuli to the sensory nerves. Probably they are all specialised epidermal cells.

Resilin The flexible cuticular protein, in the joints of arthropods.

Reticulin Fine dermal protein fibres; probably a form of collagen.

Scale keratinisation This occurs in reptilian scale epidermis, in the scaly tarsus of birds, and in rodent tail scales. It resembles parakeratosis but nuclear basophilia is lost.

Sclerotins (*scleroproteins*) Fibroin and collagen fibres bonded together by catechols or by quinones attached to the protein chains through amino and SH groups.

Septate desmosomes Characteristic tight junctions of invertebrate epidermis.

Stratum basalis Official term for the epidermal germinal basal layer, where cell division largely occurs.

Stratum corneum Official term for the vertebrate epidermal horny layer.

Stratum spinosum Official term for the vertebrate epidermal prickle cell layer.

Telogen Resting stage of hair or feather follicle.

Tight junction (*zonula occludens*) These are formed by membrane fusion between neighbouring epidermal cells.

INDEX

Acetylcholine 81
Actin filaments 17, 60
Active transport, mechanism of 172, 173
Adaptation to life on land 83, 173, 174
Adenosine monophosphate (cyclic AMP) 148, 158, 186
  triphosphatase 108, 172, 173, 187
  triphosphate (ATP) 148, 158, 161, 172
Adenyl cyclase 158, 186
Adipose tissue 105 (*see also* Subcutaneous tissue)
Adrenalin, on apocrine glands 144, 145
  on chromatophores 157
Aldosterone (*see also* Mineralocorticoids) 155
Allergic reactions 178
Ameloblasts 107
Amia 66
Amphioxus 60, 61, 62
Ampullary pits 70
Anagen (growth) stage, of feather follicle 152
  of hair follicle 127, 129, 153
Androgens (adrenocortical) 155
Annelid worms, chitin in setae 39
  collagenous cuticle 39
  quinone-bonded epicuticle 39
  tube in polychaetes 40
Antibodies 177 (*see also* Immunoglobulins)
Antidiuretic (neurohypophyseal) hormone 155, 173
Antlers 132
Anura (Amphibia) 73, 74, 77, 151
Apolysis 43
Aquatic mammals 105, 125
Arachnida 47
Archaeopteryx 91
Arrector muscles 111, 115, 128, 144, 146
Arrow worms 28, 38
Arthropods 42
  cuticular sclerotins and chitin 42
  endophragmal skeleton 42
Autograft 179
Autoradiography 8

Baldness, male pattern type in man 2, 154
Baleen 133
Basal lamina 16, 60, 65, 104, 168
Beak, of birds 94
  of cephalopods 53
Biochemical procedures 8, 9
Bioluminescence 57
Biophysical procedures 8, 9
Blood capillaries in epidermis, in Amphibia 74
  in leeches 40
  in tunicates 53, 54
Blood vessels in dermis, in Amphibia 79
  in birds 100
  in fish 65, 68
  in mammals 109
  in reptiles 90, 97
  value of vasodilation for heat loss 145
Body temperature, in birds 143
  in mammals 144
  in poikilothermic animals 141
Brachiopods 36
Bradykinin 109, 145
Breeding tubercles of fish 69, 150
Byssus 52

Calcification, in brachiopod shells 36
  in corals 23
  in crustacean cuticles, 46
  in dermal scales and bone 62, 63, 67, 106, 107
  in echinoderms 40
  in Ectoprocta 36
  in keratin 94, 132, 133

Calcification (*cont.*)
in mollusc shells 50, 51, 52
in Porifera 21
in Protozoa 20
Cancellous bone, in fish dermal scales 62, 64, 65, 71
in mammals 107
Carotenoids 56
Cartilage 48, 67, 106
Catagen 127, 128
Catechols 45, 165
Cell division 148
Cell junctions
fused junctions 76, 77, 171, 173
prickle type desmosomes 59, 121
septate desmosomes 17, 18, 170, 172
Cell migration in embryo 182
Cellulose 53, 166
Cephalopods 53, 56
Chalones 148, 186
Chameleon 89, 157
Chemical analysis 8
Chitin, chemistry 6, 8, 166
occurrence 6, 23, 36, 38, 39, 42, 52
Chitinase 8, 166
Chondrocytes 106
Chromatophores 56
control of 156, 157, 158
Chylomicrons 175
Cilia 18, 19, 31, 49
Claws 77, 84, 93, 132, 133
Clines 143
Coelacanths 70
Collagen, composition of 167, 168
in dermis 15, 102, 103
in epidermal cuticles 31, 33, 39, 162, 166
Conchiolin 51
Contact organs of fish 69
Corallite 22
Corpora allata 149
Corpora cardiaca 149
Corticosteroids (*see* Androgens, Glucocorticoids and Mineralocorticoids)
Cosmoid dermal scales 71
Crustacea 46
Cryostat in histochemistry 5
Ctenophores 25
Cuticle (epidermal) 15, 171
chemical composition of 162
ecdysis of 149, 150
in Amphibia 74
in annelids 68
in arthropods 42–8
in fish 68
in nematodes 31–5
possible remnant of, in mammals 122, 162, 182
Cyclostomes 61
Cysteine 47, 76, 87, 98, 126, 168
Cystine, invertebrate 31, 50
in keratin 87, 117, 120, 126, 164

Dehydrogenases 7, 117, 122, 124
Delayed hypersensitivity reactions 178
Dentine
in fish denticles 63, 64
in mammalian teeth 107
Dermal bony scales
in Amphibia 79
in fish 62–6, 71
in mammals 106
in reptiles 88
Dermal papilla 95, 96, 115, 184
Dermatitis 179
Dermis
amphibian 79
avian 100
elastin in 103
fish 66
fixation artefacts in mammals 101, 104
ground substance 103
invertebrate 15, 40
mammalian elastic type 101
mammalian rigid type 102
Dermo-epidermal interactions 184
Differentiation 181
DNA 159, 160, 186
Dopa reaction 3, 166

Ecdysis of cuticle
in arthropods 43, 47
in fish 68
in nematodes 33
hormonal control 149
Ecdysone 149, 158
Echinoderms 40, 42
Ectoprocta 36
Elasmobranchs 63
Elastin 16, 103, 168
Electric organs 69

Electron microscopy 7, 8
Elephant seal 125
Enamel
  in fish scales 63, 64
  in vertebrate teeth 107
Entoprocta 35
Enzymes, histochemical methods for 7
Eosinophilic cells 87
Epidermis
  growth patterns in mammals 114
  intercellular pathway 176
  mammalian types 125
  methods for separation from dermis 9
  participation in dermal scales 67
  primitive type in flatworms (Acoela) 26
  protein synthesis 116
  showing active transport 77, 171, 172
  specialisation in parasitic worms 28
  syncytium with insunken nuclei 27
  types of organisation 27
Epitheliomuscular cells 17, 22, 26
Eyes, epithelial covering 124

Feather
  follicle development 95
  growth 97, 99, 153
  keratinisation 98
  moult 97, 152
Fibrocytes 105
Fibroin proteins 162
Figures of Eberth 79
Flagella 18, 19
Flavines 57
Foetal mammalian epidermis 182
Frogs and toads 73, 151, 173

Ganoid dermal scales 65
Ganoin 64, 65
Gene activation and inactivation 186, 187
Gill epithelium of fish 69, 171, 172
Glands (compound)
  amphibian alveolar 75, 78
  apocrine 136
  arthropod silk 48
  avian preen 95, 144, 188
  eccrine sweat 134, 135
  hormonal control 155
  in molluscs 49
  in reptiles 88
  sebaceous 134
Glucocorticoids, on hair growth 154, 155
Glycoproteins 167
Golgi bodies, in protein synthesis 162
Graft rejection 179

Hagfish 61
Hair
  cycles of growth 129, 153, 154
  commercial uses 130, 131
  form 128, 129, 154
  growth 126–8, 154, 155
  moult 128, 129, 153
  phylogeny 89, 113
  structure 126
  types 112, 113
Haptens 178
Heatstroke 145
Hemichordata (Enteropneusta) 27, 53
Heparin 108
Histamine 25, 108
Histiocytes (*see* Macrophages)
Histochemical methods 6
Histological methods 5
Homografts 179
Homoiothermic animals 142
Homology 187
Hooves 132
Hormonal control, of arthropod ecdysis 149, 150
  of breeding tubercles in fish 150, 151
  of feather growth 153
  of hair growth 154
  of moult in Amphibia 151
  of moult in reptiles 152
  of sebaceous glands 155
Horns, mammalian 131, 132
Horny teeth, in cyclostomes 62
  in larval Amphibia 74
Hydra 22
Hydrolytic enzymes 76, 84, 118, 162
5-hydroxytryptamine 25, 108
Hyperkeratosis 124
Hypodermis (*see* Subcutaneous tissue)

Immunoglobulins 177
Insects 43–6
Iodine 22, 53
Iridophores 71, 80, 88

Jawless fish 62

Keratin 17
  chemistry of 161, 164, 169
  feather 99
  mammalian 117, 127, 128
  reptilian 85
Keratinisation
  cyclical in reptiles 86, 87
  in Amphibia 75
  in avian stratum corneum 92, 93
  in feather follicle 98
  in fish 69
  in hair follicle 126
  in mammalian epidermis 118
Keratinocyte retention 85
Keratohyalin 119

Lamellar (compact) bone, in fish dermal scales 62, 64
  in mammalian dermis 107
Langerhans cells 108
Lateralis system 70
Leather 104
Leeches 40, 56
Lepidosteus 66
Leydig cells 73
Lung fish 70
Lymphatics 60, 79, 110
Lymphocytes, in immune response 178
Lysosomes 118, 162

Macrophages (histiocytes) 16, 108, 180
Mantle
  brachiopod 37
  mollusc 49, 50
Mast cells 108
Meissner corpuscles 110
Melanin
  composition 163–6
  invertebrate 55
  vertebrate 71, 80, 81, 99, 109
Melanocytes 55, 80, 89, 109, 186
Melanocyte stimulating hormone (MSH)
  on melanogenesis 158
  on melanophores 157
Melanogenesis 55, 158, 163
Melanophores (*see also* Chromatophores), 55, 71, 81, 89, 109, 156–8, 166
Melanosomes 55, 71, 89, 100, 109, 157, 163, 166
Melatonin, on melanophores 157
Mesenchyme 15, 182
Mesogloea 15, 21
Methodology 11
Microvilli
  absorptive 15, 29, 175
  exosecretory 15, 17, 45, 68, 163
Mineralocorticoids
  on eccrine sweat glands 145, 155
  on moult in Amphibia (Anura) 151
Mitochondria 74, 117, 122
Modulation 183–185
Mollusc
  chitin 52
  mantle epidermis 49
  pearl formation 51
  periostracum 51
  shell calcification 52
  shell formation 50
Mother-of-pearl 51
Moulting fluid 34, 45
Mucopolysaccharides 31, 39, 103, 104, 166
Mucous goblet cells 16, 27
Mucous secretion 17, 61, 78, 167
Mucus (mucin) 17, 22, 167
Muscle in dermis 49, 79, 111
Myriapods 48

Nacreous layer 51
Nails of primates 132
Nematocysts
  development 24
  origin in flatworms 25
Nematode
  cuticle 31
  growth and ecdysis 33
Nemertina 31
Neotenin 149
Nerves to skin
  in Amphibia 79
  in birds 100
  in fish 69, 70
  in invertebrates 19, 46, 56
  in mammals, 110, 136, 145, 157
  in reptiles 89
Neural crest 55, 182
Neural sinus gland 150
Neuromast organs 70, 80

Odontocytes 107

Oestrogen
on feather growth 152, 153
on fish breeding tubercles 150, 151
on hair growth 154, 155
on sebaceous glands 134, 155
Ommochromes 56
Oral cavity 61, 90, 124
Organiser, embryonic 183
Ossification, dermal 62, 63, 71, 79, 88, 106
Osteoclasts 107
Osteocytes 106

Pacinian corpuscles 110
Pangolin scales, nature of 132
Parabiosis 149
Parakeratosis 123, 125
Pearls 51
Pellicle of Ciliata 20
Phagocytosis
by epidermal cells 157
by macrophages 16, 77, 108, 180
Phosphatases (acid and alkaline) 23, 51, 93, 108, 118, 187
Phospholipids in keratinised cells 69, 76, 87, 92, 93, 119, 120, 123
Photoperiodic pathway through eyes, importance of
in active colour change 156–8
in feather replacement 152
in hair replacement 154
in invertebrate ecdysis 150
Physiological (active) colour change
control of 156, 157
in Amphibia 81
in cephalopods and leeches 56
in fish 72
in reptiles 89
Pinocytosis by epidermal cells 175
Pituitary hormones 151, 152, 154, 155, 157, 173
Placoid denticles 63–5
Platyhelminthes 25
Pogonophora 40
Poikilothermic animals 142
Polypterus 65
Porphyrins 57
Priapulida 35
Prostaglandins and fat mobilisation 105
Proteins
chemical bonds in 161, 164
molecular structure 161, 167–9
Protein synthesis 86, 116, 159, 161, 167
DNA and 159
RNA and 160
Prothoracic glands 149
Protozoa 20
Pterines 57

Quinone linkages in sclerotins 45, 163–5

Resilin 43, 47, 48, 162
Respiratory surface, skin as 74, 176
Rhabdites 26, 31
RNA 160, 161
Rotifera 34

Scale patterns in reptiles 83, 185
Sclerotins 33, 42, 51, 56, 163
Sensation, theories of 110
Sensory receptors 19, 46, 69, 80, 89, 100, 110, 142, 145
Shell
brachiopod 37
mollusc 50
Silica
in Protozoa 20, 21
in sponges 21
Silk 48, 167
Skin colour 71, 80, 88, 89, 99, 129, 185, 186
Skin temperature in man 145
Sloughing mechanism
in birds 94
in elephant seal 125
in large flakes 125
in lizards and snakes 86, 87, 152
in tuatara 88
Sodium pump (*see also* Active transport) 172
Sponges 21
Spongin 21
Stratum corneum 75, 76, 83, 92, 93, 113, 118, 121, 122, 124
Sturgeon 66
Subcutaneous tissue (hypodermis) 60, 67, 79, 100, 105, 146
Sweat centre in hypothalamus 145
Sweating mechanism for thermal regulation 145

Teeth
development 107
types of 90

Tegument of parasitic flatworms 30
Teleost fish
  scale growth and composition 67
  value of scales in fisheries research 67
Telogen
  in feather follicle 99
  in hair follicle 128, 153, 154
Testosterone
  on feather follicles 152
  on fish breeding tubercles 151
  on hair follicles 154, 155
  on sebaceous glands 155
Thermal insulation
  by fur 146
  by plumage 144
Thermal regulation
  in desert mammals 146
  in insects 142
  in man 145
Thermosensitive centre in hypothalamus 142
Thorny-headed worms 34
Thymidine (H^3-labelled) 148
Thyroid hormone (thyroxine)
  on amphibian metamorphoris 151, 184
  on chromatophores 157
  on feather growth 152
  on hair growth 154
  on lizard and snake epidermis 152

Tissue culture 8
Tortoises and turtles 85
Tortoise-shell of commerce 85
Transpiration 173, 175
Trichohyalin 126
Tropocollagen 167
Tunicate test (tunic) 53, 54
Tyrosinase 109, 166

Urodela (Amphibia) 73, 74

Vasodilation 144
Velvet in deer 132
Vibrissae 112, 126
Vitamin A (retinol), effects on epidermis 10, 185

Water
  diffusion barrier in insect epicuticle 173
  diffusion barrier in vertebrate stratum corneum 174
  storage in skin 103, 109
Waxes in arthropods 43, 173
Wool 130

Xanthophores (lipophores) 71, 81, 89
Xenograft 177

Y organ 150

D0271966